Symbolic Systems Biology

Jones & Bartlett Learning Titles in Biological Science

AIDS: Science and Society, Sixth Edition
Hung Fan, Ross F. Conner, & Luis P. Villarreal

AIDS: The Biological Basis, Fifth Edition
Benjamin S. Weeks & I. Edward Alcamo

Alcamo's Fundamentals of Microbiology: Body Systems Edition
Jeffrey C. Pommerville

Alcamo's Fundamentals of Microbiology, Ninth Edition
Jeffrey C. Pommerville

Aquatic Entomology
W. Patrick McCafferty & Arwin V. Provonsha

Bioethics: An Introduction to the History, Methods, and Practice, Second Edition
Nancy S. Jecker, Albert R. Jonsen, & Robert A. Pearlman

Bioimaging: Current Concepts in Light and Electron Microscopy
Douglas E. Chandler & Robert W. Roberson

Biomedical Graduate School: A Planning Guide to the Admissions Process
David J. McKean & Ted R. Johnson

Biomedical Informatics: A Data User's Guide
Jules J. Berman

Botany: An Introduction to Plant Biology, Fourth Edition
James D. Mauseth

Case Studies for Understanding the Human Body
Stanton Braude, Deena Goran, & Shelley Maxfield

Clinical Information Systems: Overcoming Adverse Consequences
Dean F. Sittig & Joan S. Ash

Defending Evolution: A Guide to the Evolution/Creation Controversy
Brian J. Alters

The Ecology of Agroecosystems
John H. Vandermeer

Electron Microscopy, Second Edition
John J. Bozzola & Lonnie D. Russell

Encounters in Microbiology, Volume 1, Second Edition
Jeffrey C. Pommerville

Encounters in Microbiology, Volume 2
Jeffrey C. Pommerville

Equilibria and Kinetics of Biological Macromolecules
Jan Hermans & Barry R. Lentz

Essential Genetics: A Genomics Perspective, Fifth Edition
Daniel L. Hartl

Essentials of Molecular Biology, Fourth Edition
George M. Malacinski

Evolution: Principles and Processes
Brian K. Hall

Exploring Bioinformatics: A Project-Based Approach
Caroline St. Clair & Jonathan E. Visick

Exploring the Way Life Works: The Science of Biology
Mahlon Hoagland, Bert Dodson, & Judy Hauck

Genetics: Analysis of Genes and Genomes, Seventh Edition
Daniel L. Hartl & Elizabeth W. Jones (late)

Genetics of Populations, Fourth Edition
Philip W. Hedrick

Guide to Infectious Diseases by Body System
Jeffrey C. Pommerville

Human Biology, Sixth Edition
Daniel D. Chiras

Human Biology Laboratory Manual
Charles Welsh

Human Embryonic Stem Cells, Second Edition
Ann A. Kiessling & Scott C. Anderson

Introduction to the Biology of Marine Life, Ninth Edition
John F. Morrissey & James L. Sumich

Laboratory and Field Investigations in Marine Life, Ninth Editi
James L. Sumich & Gordon Dudley

Laboratory Fundamentals of Microbiology, Ninth Edition
Jeffrey C. Pommerville

Laboratory Investigations in Molecular Biology
Steven A. Williams, Barton E. Slatko, & John R. McCarrey

Laboratory Textbook of Anatomy and Physiology: Cat Versi Ninth Edition
Anne B. Donnersberger

Lewin's CELLS, Second Edition
Lynne Cassimeris, Vishwanath R. Lingappa, & George Plop

Lewin's Essential GENES, Second Edition
Jocelyn E. Krebs, Elliott S. Goldstein, & Stephen T. Kilpatr

Lewin's GENES X
Jocelyn E. Krebs, Elliott S. Goldstein, & Stephen T. Kilpatr

Mammalogy, Fifth Edition
Terry A. Vaughan, James M. Ryan, & Nicholas J. Czaplews

Microbes and Society, Second Edition
Benjamin S. Weeks & I. Edward Alcamo

The Microbial Challenge: Science, Disease, and Public Hea Second Edition
Robert I. Krasner

Microbial Genetics, Second Edition
Stanley R. Maloy, John E. Cronan, Jr., & David Freifelder

Microbiology Pearls of Wisdom, Second Edition
S. James Booth

Molecular Biology: Genes to Proteins, Third Edition
Burton E. Tropp

Neoplasms: Principles of Development and Diversity
Jules J. Berman

Perl Programming for Medicine and Biology
Jules J. Berman

Plant Cell Biology
Brian E. S. Gunning & Martin W. Steer

Plants, Genes, and Crop Biotechnology, Second Edition
Maarten J. Chrispeels & David E. Sadava

Plant Structure: A Color Guide, Second Edition
Bryan G. Bowes & James D. Mauseth

Precancer: The Beginning and the End of Cancer
Jules J. Berman

Principles of Modern Microbiology
Mark Wheelis

Protein Microarrays
Mark Schena, ed.

Python for Bioinformatics
Jason Kinser

R for Medicine and Biology
Paul D. Lewis

Ruby Programming for Medicine and Biology
Jules J. Berman

Strickberger's Evolution, Fourth Edition
Brian K. Hall & Benedikt Hallgrímsson

Tropical Forests
Bernard A. Marcus

20th Century Microbe Hunters
Robert I. Krasner

Understanding Viruses
Teri Shors

Symbolic Systems Biology

Theory and Methods

Edited by
M. Sriram Iyengar, PhD
Assistant Professor
School of Health Information Sciences
University of Texas Health Science Center at Houston
Houston, Texas

JONES AND BARTLETT LEARNING
Sudbury, Massachusetts
BOSTON TORONTO LONDON SINGAPORE

World Headquarters
Jones & Bartlett Learning
40 Tall Pine Drive
Sudbury, MA 01776
978-443-5000
info@jblearning.com
www.jblearning.com

Jones & Bartlett Learning Canada
6339 Ormindale Way
Mississauga, Ontario L5V 1J2
Canada

Jones & Bartlett Learning International
Barb House, Barb Mews
London W6 7PA
United Kingdom

Jones & Bartlett Learning books and products are available through most bookstores and online booksellers. To contact Jones & Bartlett Learning directly, call 800-832-0034, fax 978-443-8000, or visit our website, www.jblearning.com.

Substantial discounts on bulk quantities of Jones & Bartlett Learning publications are available to corporations, professional associations, and other qualified organizations. For details and specific discount information, contact the special sales department at Jones & Bartlett Learning via the above contact information or send an email to specialsales@jblearning.com.

Production Credits
Publisher, Higher Education: Cathleen Sether
Acquisitions Editor: Molly Steinbach
Senior Editorial Assistant: Jessica Acox
Editorial Assistant: Caroline Perry
Production Manager: Louis C. Bruno, Jr.
Senior Production Editor: Katherine Crighton
Senior Marketing Manager: Andrea DeFronzo
V.P., Manufacturing and Inventory Control: Therese Connell
Composition: Lapiz
Cover Design: Kristin E. Parker
Photo Research and Permissions Manager: Kimberly Potvin
Cover Image: © Sebastian Kaulitzki/ShutterStock, Inc.
Printing and Binding: Malloy, Inc.
Cover Printing: John Pow Company

Library of Congress Cataloging-in-Publication Data
Symbolic systems biology : theory and methods / edited by M. Sriram Iyengar.
p. cm.
Includes bibliographical references and index.
ISBN-13: 978-0-7637-5370-2 (cloth)
ISBN-10: 0-7637-5370-X (cloth)
1. Systems biology. I. Iyengar, M. Sriram.
QH324.2.S946 2009
572–dc22 2009051932

6048

Printed in the United States of America
14 13 12 11 10 10 9 8 7 6 5 4 3 2 1

Contents

Preface

The Astonishing Complexity of the Living World

A revolution is unfolding in the biological sciences. The rate and scale of biological data being gathered with new microbiological technologies is rapidly outpacing classic methods of data analysis. At the same time, it is becoming increasingly apparent that simple linear one-cause-one-effect relationships are the exception, not the rule in many complex processes in living organisms. These trends highlight the astonishing complexity of the living world, and demand new techniques to help working biologists answer the most important problems of the day.

In the last decade, Systems Biology (SB), the mathematical and computational modeling of complex biological processes, has provided increasingly important leverage for biology. The underlying motivation is to gain deeper insights into complex biological phenomena that could enable development of meaningful predictions and help biomedical scientists to design and perform focused wet-lab experiments. Such focused experiments can lead to faster and cheaper development of therapies for diseases including cancer, cardiovascular disease, neural disorders, and others.

Historically, mathematics and, more recently, computing, have been the principal tools for systematic understanding of inherently quantitative sciences such as physics and engineering. The major obstacle to their fulfilling a similar role in biology and medicine is that these disciplines have typically been regarded as largely descriptive and fraught with immense complexity, the extent of which is still being elucidated.

It is an understatement to say that biology is complex. The human body consists of a symbiotic community of an estimated 100 trillion living cells, classified into many species of commensal organisms (concentrated in the gut and skin) and about 500 types of human cells. Each cell performs specific functionality in addition to the expression of up to 30,000 distinct proteins, energy transformations, electrical and biochemical signaling, division, and proliferation, among others. Biologists handle all this complexity by concentrating on specialties and sub-sub-specialties. Yet each of these specialties can be characterized by staggering complexity. Consider neurobiology, a principal focus of which is the neuron, a cell specialized for transmitting electrochemical signals. A cubic centimeter of the human brain can contain 50 million neurons, each of which is associated with 6 to 10 glial cells and an unknown number of microglial cells. A human neuron can make multiple connections to thousands of other neurons, so that the brain contains tens of trillions of interconnections between neurons. The mere contemplation of all of this is enough to astonish even the most jaded computer architect.

This catalog of lifes complexity is boundless and exhausting. Here, it suffices to summarize by saying that at any instant the human body is the venue for thousands of trillions of interacting events, each modulated by a host of factors including the physical environment, genetic expression, and by exogeneous and endogeneous stimuli, including those of psychosomatic origin. On top of all this is another layer of complexity in the form of stochastic effects, caused by the inescapable and ubiquitous presence of randomness in all biological processes.

Given this state of affairs, the current high level of interest in Systems Biology (SB) is understandable. It holds the exciting promise of transitioning biology from a compendium of facts to a discipline supported closely by mathematical and computing frameworks. This is not to say that biology lacks theoretical underpinnings. The central dogma of molecular biology (DNA to RNA to Protein), laws of genetics, and other theories provide predictive logic and underlying explanatory theory. However, current SB is concerned with layers of biological complexity beyond the genome. The focus of much of SB research presently is the biological cell, specifically on interactions between intracellular biomolecules, mostly proteins, comprising metabolic and signaling pathways and networks. One motivation for this has its origins in the year 2000 upon the completion of the human genome project and the surprising result that the 3 billion base-pairs in human DNA contain only about 30,000 genes while lowly organisms such as earthworms possess about 25,000. Thus, the vast differences in almost every respect between the two classes of organisms must be caused by extra-genomic factors, perhaps in the millions of proteins significant in biology and collectively called the proteome.

Mathematical and Computational Modeling

Until the widespread availability of computing in the early 1970s most analytic and modeling techniques for science and technology had their origins in mathematics. Significant examples for biology include the Michaelis-Menten equations for describing enzyme kinetics, the McCulloch-Pitts model for neuron activity, Turings studies in morphogenesis, the Nobel-prizewinning work of Hogkins and Huxley describing membrane potentials in neurophysiology, and Rene Thom and Christopher Zeemans abstract models of structural stability based on the formers Catastrophe Theory. The 1960s also saw considerable interest in what was then called Theoretical Biology under the leadership of the noted geneticist Conrad Waddington. Theoretical Biology may be viewed as the direct ancestor of present-day SB. However, these mathematical techniques, based often on differential equations, were computationally difficult to extend beyond small numbers of components. Perhaps more significant was that the rigorous structure of these techniques, based in part on the quantitative idealizations of calculus, was found to be inadequate to model the multifaceted complexity of biological processes. For example, biological pathways are not governed by quantitative factors alone. Qualitative aspects such as the intracellular location of the protein-protein reaction of interest play crucial roles.

The steadily increasing availability of powerful digital computers after the mid 1970s enabled new methods of modeling, based on algorithmic computation rather than pure mathematics. The difference between computational and mathematical modeling is that the former can implement and execute a much wider spectrum of reasoning at multiple layers of abstraction than that permitted by the quantitative formalisms of calculus. Other benefits of computational models include the ability to search very rapidly through vast troves of data, and to execute complicated algorithms consisting of tens of thousands of mathematical and logical steps in microseconds. In the present day, the role and applicability of computational modeling has been greatly enhanced by the increasing availability of affordable and very powerful computing hardware as well as easy-to-use software development environments on desktops.

Two Schools of Systems Biology

A principal focus area of present-day SB is to analyze and develop predictive models of intracellular protein-protein interactions. Such interactions, in the hundreds or thousands, form the components of complex intertwining networks classified into two broad areas: *metabolic and signal-transduction* pathways. Broadly speaking, the former is concerned with energy production and usage within the cell, while the latter is responsible for transitions to cellular states such as apoptosis, cell death, and proliferation, among others. Both kinds of pathways can be initiated by extracellular events, such as the arrival of signaling molecules on the surface of the cell, and modulated by genomic processes within the cell. Clearly, in-depth understanding of these pathways is of great importance not only to biology, but also to medicine. For example, a thorough understanding of the signaling processes which cause cell proliferation could lead to therapies for cancer.

Systems of simultaneous differential equations based on the well-developed framework of reaction kinetics and thermodynamics were initially developed to model these pathways. However, certain difficulties with these techniques became apparent in the mid-1990s. This quantitative approach entails detailed knowledge of rates, concentrations, 3-D structural details, and other parameters of the equations. Meaningful insights into the behavior of networks can be derived only if accurate data are available for all the coefficients involved. The problem is that measuring accurate rates and/or concentrations of intracellular biomolecules is a daunting if not impossible task, especially given the possibility of stochastic effects. Concentrations that can make a difference range down to picograms or, in extreme cases, single molecules, and can often not be distinguished from sensor noise in current biochemical assays. However, the stability of solutions of differential equations is often crucially dependent on the values of these parameters in the sense that slight variations in input parameters may result in large variations in solution values. Further, these systems of simultaneous equations can become computationally intractable to analyze using those tools if they involve more than a few tens of nodes (proteins) in the pathway being analyzed. Thus, the scalability of these quantitative models to networks involving hundreds of proteins requires careful consideration and further research. Interestingly, similar difficulties have recently been identified with respect to the use of Hodgkin-Huxley models in neurobiology.

Faced with these difficulties, systems biologists in the mid-1990s began devising alternative analytic and modeling strategies based on computational rather than mathematical machinery. The view here is that the thousands of biochemicals interacting in signaling or metabolic pathways within a cell are performing a species of concurrent computation. The challenge is to devise appropriate algorithmic models that reflect the known nature and behavior of biological pathways, and that can be computationally tractable at the level encountered in biology. Naturally, these *in silico* models should also enable rigorous, experimentally verifiable prediction and simulation. As it happens, developing rigorous, analytical models of complex computations, including concurrent, parallel, and distributed systems, has been (and remains) a topic of intense interest to computer scientists.

Typically, when such models are applied in biology, the components of biological pathways (proteins, for example), are represented as discrete symbols, together with algebraic rules governing the interactions between these components. In this strategy, qualitative factors such as compartmentalization and localization within the cell (plasma membrane, cytoplasm, nucleus, etc.) of protein-protein reactions can be represented as attributes of the symbols involved. These kinds of strategies have been very successfully applied in areas such as network and chip design, since they enable productive reasoning of the behavior of these structures at several levels of abstraction, while not being constrained by having to know exact values of thermodynamic and kinetic coefficients. In other words, designing the placement of transistors, resistors, diodes, gates, and other components of an integrated circuit does not require knowledge of subatomic processes and parameters governing movement of electrons.

Correspondingly, in the case of signaling or metabolic pathways, at a certain level of abstraction, it is sufficient to know that protein X can exist in inactivated or activated states. In an activated state it has either an inhibitory or excitatory effect on protein Y. In turn, protein Y, when activated, has an inhibitory or excitatory (or other) effect on protein Z, and so on. Thus, there is no need to delve deeply into the thermodynamics of the interactions between proteins X and Y, or the hundreds of others potentially participating in a network, to reason formally, accurately, and productively. The dynamics of the pathway can be represented as state transitions and the conditions under which these occur can be defined using algebraic rules. In this way, the system can scale to complex pathways involving hundreds of components. If needed, attributes can be added to the symbols to take the model to a finer level of abstraction. Such additional attributes can include intra-protein structures including specific sequences of amino acids such as protein functional domains. Models at these finer levels can correspondingly yield greater insights.

Symbolic Systems Biology

The latter approach has come to be known as *Symbolic Systems Biology* and is the focus of many groups of systems biologists worldwide. It is becoming recognized as complementary to the strictly quantitative, mathematically oriented approaches, capable of answering different kinds of questions at several levels of abstraction. In a sense, this approach is not entirely new; it can be argued that the Linnaean system is a symbolic ontology of the living world. The power of abstraction can also be observed in Watson and Cricks use of symbolic models of the 3-D geometry of nucleic acids to reach the crucial insights that led to the elucidation of the structure of DNA. The reasoning they followed did not need to take into account the thermodynamics of chemical bonding, or require knowledge of corresponding reaction rate constants, to reach a conclusion that has profoundly influenced biology.

As in any rapidly developing area, SSB researchers are exploring multiple approaches, drawing upon existing formalisms for abstract and concurrent computation that have proven to be useful in other (non-biological) areas. Multiple tools, sometimes overlapping in intent, are under various stages of development, from laboratory prototypes to well-developed and well-supported commercial products. This plurality of methods is one of the strengths of the symbolic approach. No one technique or methodology can hope to satisfy the entire space of questions posed by biologists, drug developers, physicians, and others engaged in biomedical research. Equally, biology is too complex to be modeled by any single paradigm.

The intent of this book is to present an overview of a selection of these techniques to students and practitioners. We have been fortunate in being able to enlist experts to write about the techniques that they have pioneered. Chapter 1 presents Pathway Logic, a formalism based on rewriting logic and implemented by the Maude system, by Patrick Lincoln and Carolyn Talcott. Chapter 2, by Maria Kwiatkowska, Gethin Norman, and Dave Parker, models stochastic behavior in biological systems using probabilistic model checking, and describes PRISM, software developed for this purpose. Chapter 3, by Monika Heiner, presents the application of Petri Nets in biology. Petri Nets, a formalism developed for modeling distributed systems, have successfully been applied in workflow modeling, concurrent computation, and others. In Chapter 4, Jasmin Fisher and David Harel present the point of view that biological systems are a class of *reactive systems* and present the Statecharts methodology. A reactive system is one that exhibits complex responses to stimuli. Chapter 5, by Andrew Phillips, applies stochastic pi calculus to help understand complex, interacting, and concurrent biological processes, and describes the SpiM simulation system. The Pi Calculus, a model of concurrent computation, has been successfully applied in diverse areas including business process modeling; the stochastic version enables the introduction of probabilistic elements. Finally Chapter 6, by Mary McGuire and the editor, M. Sriram Iyengar, is a sampling of current symbolic systems biology tools and methodologies. It is noteworthy that the models presented in Chapters 1 through 5 identify all or some of concurrency, interactivity among multiple executing processes, stochastic behavior, and reactivity, as characteristic of biological systems and processes.

The reader can select one or a combination of several of the presented approaches to apply to a biological problem of interest. Important topics have undoubtedly been omitted, and for this the editor apologizes, pleading the necessity of producing a volume of manageable size. Also, the increasing numbers of researchers in this field and pace of innovations mean that this volume is at best a snapshot of some approaches at one point in time.

Interested readers are pointed toward leading journals, conferences, and web sites to view the latest and greatest advances in this exciting area of research.

M. Sriram Iyengar, Patrick Lincoln

Publisher's Note

Color versions of many of the figures in Symbolic Systems Biology can be found on the book's website, at http://www.jblearning.com/science/molecular.

Chapter

1

Symbolic Systems Biology and Pathway Logic

Patrick Lincoln and Carolyn Talcott

SRI International
333 Ravenswood Avenue, Menlo Park, CA 94025

Abstract Pathway Logic (PL) is an approach to modeling and analysis of biological processes based on rewriting logic. This chapter describes the use of PL to model signal transduction processes. It begins with a general discussion of symbolic systems biology, followed by some background on rewriting logic and executable specifications. Finally, representation and analysis of a small model of Ras and Raf activation is presented in some detail.

Key words: Symbolic systems biology, rewriting logic, signal transduction, Pathway Logic

1.1 Symbolic Modeling of Cellular Processes

Biological processes are complex. They exhibit dynamics with a huge range of time scales: microseconds to years. The spatial scales cover 12 orders of magnitude: metabolite to single protein to cell to organ to whole organism. Just at the cellular level, cells interact with their environment, both sensing and affecting. They have many behaviors: They can grow, proliferate, migrate, differentiate, or die. Underlying these behaviors are a variety of processes, such as gene regulation, signal transduction, and metabolism, that interact with one another in complex ways. Genes are regulated by proteins (and other molecular entities) that bind to promoter regions. This binding determines which genes are expressed (turned on) and, therefore, which new proteins are produced. These proteins may, in turn, regulate the same or other genes. A cell senses its environment through receptors in the membrane that recognize specific types of molecules or conditions. This recognition results in *signal transduction* that transmits the information to appropriate components inside the cell. Mechanisms underlying the flow of information include modification of protein state, formation of complexes, and change of location. The flow is controlled by mechanisms that either activate or inactivate proteins in the signaling path. Metabolism involves both synthesis and degradation of chemicals to generate energy, synthesize protein building blocks (amino acids), and cell structure components, among other actions. Metabolic processes are controlled by enzymes, which may in turn be activated or inhibited by signaling processes. Furthermore, metabolites such as glucose play a role in controlling signal flow.

Oceans of experimental biological data are being generated, from both traditional and emerging high-throughput techniques. How can we use these data to develop better models? Important intuitions are captured in mental models of biological processes that biologists build and the cartoons they draw. The trouble is that these models are not amenable to computational analysis.

High-level statistical models can be developed, for example, to discover possible correlations and causal relations. Such models may suggest useful insights, but have many limitations, such as features that cannot be modeled. Low-level, detailed kinetic or stochastic models can be developed for small subsystems, but often require parameter fitting, so that the reaction rates used reflect unknown biological context.

Symbolic systems biology is the *qualitative* and *quantitative* study of biological processes as integrated systems rather than as isolated parts. Our focus here is on modeling causal networks of biomolecular interactions in a logical framework at multiple scales. The aim is to develop formal models that are as close as possible to domain experts (biologists) and mental models. Furthermore, it is important to be able to perform calculations with and analyze these complex networks. The latter endeavor includes techniques for abstracting and refining the logical models; using simulation and deduction to compute or check postulated properties; and making testable predictions about possible outcomes, using experimental results to update the models.

Many challenges arise in the course of developing symbolic systems models. One challenge is choosing the right abstractions. Biological networks (metabolic, protein, or regulatory, for example) are large and diverse.

It is important to balance computational complexity against model fidelity and to be able to move between models of different levels of detail, using different formalisms in meaningful ways. Biological networks can be combined to produce high levels of physiological organization; for example, circadian clock subnetworks are integrated with metabolic, survival, and growth subnetworks. A second challenge is to compose different views or models of different components into integrated system models.

Symbolic/logical models allow one to represent partial information and to model and analyze systems at multiple levels of detail, depending on the information available and the questions to be studied. Such models are based on formalisms that provide language for representing system states and mechanisms of change such as reactions. These languages come with well-defined semantics, and tools for analysis are based on these underlying semantics. Formal theories can include both specific facts and general principles relating and categorizing data elements and processes. New data structures for representing biological entities and their relations and properties can easily be defined. Theories concerning different types of information can also be combined using well-understood operations for combining logical theories. A wide range of analytical tools developed for the analysis of computer system specifications is being adapted to carry out new kinds of analysis of experimental data curated into formal theories.

Of particular interest are symbolic models that are *executable*. An executable model describes system states and provides rules specifying the ways in which the state may change. Such models can be used for simulation of system behavior. In addition, properties of processes can be stated in associated logical languages and checked using tools for formal analysis. Given an executable model, the path graph of a given initial state is a graph whose nodes are the reachable states and whose edges are the rules connecting them. Paths through the graph then correspond to possible ways in which a system can evolve. An execution strategy picks out a particular path among the various options possible. For such a model, many kinds of analysis can be carried out, including static analysis, forward simulation, forward search, backward search, model checking, constraint solving, and meta-analysis.

Static analysis allows one to examine the structure of the model and to understand how the elements are related and organized (the sort structure). It can be used to infer flow of control and dependencies. Static analysis also provides a means to check for inconsistencies or ill-formed declarations and to look for missing information.

Forward simulation runs the model from a given initial state using a specified strategy, either for a fixed number of steps or until no more rewrites apply. This type of processing is extremely fast, and very useful for initial exploration.

Forward search is a breadth-first search of all paths through the transition graph for a given initial state. If the graph (number of states) is finite, such a search will find *all* possible outcomes from a given initial state. Assuming finite branching, the search will find all outcomes at a finite depth. A search can also be constrained to find only those states satisfying a given property.

Backward search runs the model backward. For models satisfying certain constraints, a backward search can answer the question: "From which initial states can we get to this state?" For example, this strategy can be used to find all possible precursors to a particular checkpoint, or to prove that a bad state cannot be reached from an initial state.

Model checking expands the collection of properties that can be investigated. Such a search concerns only properties of individual states. Model-checking tools are based on algorithms to determine if all computations of a system (pathways, sequences of steps) satisfy a given property. For example, we can ask if molecule X is never produced before molecule Y has been produced. If not, a pathway that fails to satisfy the property (molecule Y is produced and molecule X is produced before it) is returned. Turning this statement around, so as to find a pathway satisfying a property of particular interest, one asserts that no such pathway exists and a counterexample will be one of the desired pathways. An example of another kind of property that can be model checked is this: "If we reach a state that satisfies P, then do we always later reach a state satisfying Q?"

Constraint solving attempts to find values for a set of variables that satisfy a given set of constraints. Maximal satisfiability (Max Sat) problems are a generalization of constraint satisfaction problems where conflicting constraints may be present, and hence no assignment of values to variables will satisfy them all. Weights (importance measures) are assigned to constraints, and a Max Sat solver finds a solution maximizing the total weight of the satisfied constraints. Many static analysis problems can be formulated as constraint systems. Steady-state analyses, such as determining possible flows of information or chemicals through a system, can be formulated as constraint problems.

Meta-analysis allows us to reason about the models themselves. Essential features of models can be abstracted to form families of related models, allowing us to work with uncertainty about reactions. Starting with a base set of known reactions, different instantiations of sets of reactions can be explored. For example, we can search for models where a given path property is true in a given initial state. In addition, rules themselves can be abstracted into families of rules, with each family corresponding, for example, to a particular type of reaction such as activation, inhibition, or translocation. This practice also allows the knowledge base to be queried as a database—for example, finding all rules that involve a given protein (in any or a specified state or location). Finally, using mappings of logics, a model can be mapped to another formalism to take advantage of additional tools.

A variety of formalisms initially developed to model and analyze concurrent computer systems have been used to develop symbolic models of biological systems, including Petri nets [14, 17, 21, 25, 28, 30, 41, 53]; the pi-calculus [32, 33, 43] and its stochastic variants [39]; membrane calculi [4, 34, 42]; Statecharts [9, 18], Live Sequence Charts [22]; rule-based systems including rewriting logic [7, 10, 11, 31, 48, 50]; BioCham [2, 5, 12]; P-systems [35, 40]; and hybrid systems [15, 19, 28, 51]. For a recent review of "executable specification approaches," see [13]. A comprehensive review of rule-based modeling can be found in [20]. Overviews of different Petri net formalisms and their application to modeling biological processes can be found in [6, 16]. A series of abstract machines, each suited to modeling biological process associated to a different class of macromolecules, is presented in [3] giving a nice introduction to the concepts to be modeled.

Symbolic executable models can be mapped to alternative logical formalisms for analysis. As will be discussed later, certain rewriting logic models can be mapped to Petri nets for analysis by special purpose, efficient model checkers. In [1], a continuous stochastic logic and the probabilistic symbolic model checker, PRISM, are used to express and check a variety of temporal queries for both transient and steady-state behaviors. Proteins are modeled as synchronous concurrent processes, and concentrations are modeled by discrete, abstract quantities. Metabolic or signaling networks can be analyzed using a constraint-based technique that generalizes the well-known flux balance analysis [8] by representing the network as constraints on the reactions, rather than as constraints on the reacting components. In [52], this technique is used to compute preferred steady states under different conditions, also represented as constraints. Apart from understanding the steady-state configurations, constraint-based analysis can also be used to identify modules in the network, trace the flow of information in the network, and identify cross-talk and conflicts.

In the remainder of this chapter, we describe the Pathway Logic approach to symbolic systems biology.

1.2 Pathway Logic Overview

Pathway Logic (PL) [10, 11, 47–50] is a symbolic systems biology approach to the modeling and analysis of molecular and cellular processes based on rewriting logic [31]. In PL, biological molecules, their states, locations, and their roles in molecular or cellular processes can be modeled at very different levels of abstraction. For example, a complex signaling protein can be modeled either according to an overall state based on its post-translational modifications, or as a collection of protein functional domains and their internal or external interactions. Similarly, biological processes can be represented at different levels of granularity using rewrite rules. Each rule represents a step (at the chosen level of granularity) in a biological process such as metabolism or intracellular/intercellular signaling. A rule may represent a family of reactions using variables to stand for families of molecular components. Rules express dependencies on biological context—for example, a scaffold needed to hold proteins in position to interact productively.

A collection of rules, together with the underlying data type specifications, forms a PL knowledge base. Each biological molecule that is declared in a PL rewrite theory has associated metadata linking it to standard database entries—for example, HUGO or UniProt/Swiss-Prot for proteins, along with other information such as category and synonyms. This information is part of the knowledge base. It is important to place the knowledge in a broader context and to be able to integrate it with other knowledge sources. Each rule has associated evidence used to justify the rule, which is also part of the knowledge base.

A PL model consists of a specification of an initial state (cell components and locations) interpreted in the context of a knowledge base. Such models are executable and can be understood as specifying possible ways a system can evolve. Logical inference and analysis techniques are used for simulation to study possible ways a system could evolve, to assemble pathways as answers to queries, and to reason about dynamic assembly of complexes, cascading transmission of signals, feedback loops, cross-talk between subsystems, and larger pathways. Logical and computational reflection are used to transform and further analyze models.

Pathways are not predefined. Instead, they are assembled by applying the rules starting from an initial state, searching for a state meeting given conditions. For example, a pathway leading to specific conditions, such as activation of a Ras protein, can be generated as the result of a logical query. A subnet (subset of reactions) composed of all possible relevant pathways can also be generated. A subnet consisting of connections to a given set of molecular components can be generated by graph exploration techniques.

PL knowledge is represented and analyzed using Maude [7], a rewriting logic-based formalism. The Pathway Logic Assistant (PLA) [49] provides an interactive visual representation of PL models. In PLA, models are represented as graphs with nodes for rules and components, and edges connecting reactant components to rules and rules to product components (formally, these graphs are known as Petri nets). These models can be queried and in silico experiments can be performed to study the effects of perturbations on these networks. Using PLA, a biologist can do the following:

- Ask for a list of dishes available for study, and modify or create dishes
- Display the network of signaling reactions for a specified model

- Formulate and submit queries to find pathways—for example, activating one protein without activating a second protein, or exhibiting a phenotype signature such as apoptosis
- Compare two pathways
- Find knockouts—proteins whose omission prevents reaching a specified state
- Incrementally explore network connections to given rules or components

PLA, sample models, tutorial material, papers, and presentations are available from the Pathway Logic website, `http://pl.csl.sri.com/`.

1.3 Introduction to Formal Executable Specification and Maude

As mentioned in Section 1.2, Pathway Logic models of biological processes are developed using the Maude system, a formal language and tool set based on rewriting logic. Rewriting logic [31] is a logical formalism that is based on two simple ideas: States of a system are represented as elements of an algebraic data type, specified in an equational theory, and the behavior of a system is given by local transitions between states described by *rewrite rules.* An equational theory specifies data types by declaring constants and constructor operations that build complex structured data from simpler parts. Functions on the specified data types are defined by *equations* that allow one to compute the result of applying the function. A term is a variable, a constant, or an application of a constructor or function symbol to a list of terms. A specific data element is represented by a term containing no variables. Assuming the equations fully define the function symbols, each data element has a canonical representation as a term containing only constants and constructors. This canonical representation is obtained by using the equations to rewrite the data term. An equation is applied to a term by matching the left-hand side of the equation to a subterm and then replacing that subterm by the corresponding right-hand side of the equation. For example, the natural numbers are constructed from the constant 0 by application of the successor function $s(0), s(s(0))\ldots$ (usually written as $1, 2, \ldots$). The plus function, $+$, can be defined by two equations: $n + 0 = n$ and $n + s(m) = s(n) + m$, where n and m are variables standing for arbitrary numbers. Using these equations, we can compute the canonical form (value) of $s(s(0)) + s(s(s(0)))$ $(2 + 3)$.

The second equation matches $s(s(0)) + s(s(s(0)))$ with $n := s(s(0))$ and $m := s(s(0))$ and the result of rewriting with this equation is $s(s(s(0))) + s(s(0))$. Using the second equation twice more, we get $s(s(s(s(0)))) + s(0)$ and then $s(s(s(s(s(0))))) + 0$. Now we apply the first equation to obtain $s(s(s(s(s(0)))))$ (i.e., 5).

One data type might be a subtype (subsort, subset) of another. For example, the non-zero numbers are a subset of all numbers. Elements of one data type might consist of lists or multisets of elements from another type. For example, a system might be represented by a set of pairs such as `{(A,2) (B,5) (C,0)}`.

A rewrite rule has the form $t \Rightarrow t'$ *if* c where t and t' are patterns (terms possibly containing variables) and c is a condition (a Boolean term). Such a rule applies to a system in state s if t can be matched to a part of s by supplying the appropriate values for the variables, and if the condition c holds when those values are used. In this case, the rule can be applied by replacing the part of s matching t by t' using the matching values for the placeholders in t'. The process of applying the rewrite rules generates computations (also thought of as deductions). In the case of biological processes, these computations correspond to pathways. Note that rewriting with rules is similar to rewriting with equations, in that we match the left-hand side and replace the matched subterm by the instantiated right-hand side. The difference lies in the way the rewriting is used. Equations are used to define functions by providing a means of computing the value of a function application. As a consequence, the equations of an equational theory should give the same result independent of the order in which they are applied. Furthermore, equational rewriting should terminate. In contrast, rules are used to describe change over time, rather than computing the value of a function. They often describe nondeterministic, possibly infinite behavior.

To summarize, a rewriting logic specification naturally has two parts: an equational part that specifies data types and functions on these types, and a rules part that specifies how systems may evolve. To query the specification, one may further specify one or more terms representing systems (initial states) of interest, to be analyzed using formal tools such as execution, search, and model checking.

Maude is a language and tool based on rewriting logic (it can be found at `http://maude.cs.uiuc.edu`). Maude provides a high-performance rewriting engine featuring matching modulo associativity, commutativity, and identity axioms, as well as search and model-checking capabilities. Thus, given a specification S of a concurrent system, one can execute S to find one possible behavior; use a search to see whether a state meeting a given condition can be reached; or model-check S to see if a temporal property is satisfied and, if it is not satisfied, to see a computation that is a counterexample.

The following simple example introduces Maude notation and gives some intuition about how to represent and analyze the structure and behavior of concurrent systems using Maude. We call this example *Magic Marbles*. In the world of magic marbles, a marble can be plain or it can have some magical potential: positive, negative, or (positive or negative) activator.

A positive (respectively, negative) activator marble can give positive (respectively, negative) potential to a plain marble. When it does so, it changes parity and becomes a negative (respectively, positive) activator. If a marble with negative potential contacts a marble with positive potential, the potential is cancelled and both marbles become plain. A marbles world consists of a collection (formally a multiset) of marbles interacting according to these laws.

We formalize the marbles world in Maude by defining three modules. The modules `MAGIC-MARBLES-DATA` and `MAGIC-MARBLES-STATE` specify data types representing marbles and marbles world states. The module `MAGIC-MARBLES-RULES` specifies the rules governing magic marble behavior. The modules `MAGIC-MARBLES-DATA` and `MAGIC-MARBLES-STATE` are functional modules specifying an equational theory. They form the equational part of the marbles world specification. The module `MAGIC-MARBLES-RULES` is a system module forming the rules part.

A Maude module begins with the keyword `fmod` (a functional module, specifying one or more data types) or `mod` (a system module, with rules specifying system behavior), followed by the module name, and ends with the corresponding keyword `endfm` or `endm`, respectively.

The module `MAGIC-MARBLES-DATA` begins by declaring a sort `Marble`, the data type consisting of all marbles, and a subsort (think subset or subtype) `PlainMarble`, consisting of plain marbles. This is followed by an `ops` declaration naming several specific plain marbles—for example, a red marble `redM`. Next, a sort `MagicMarble` of marbles with magical potential is declared. It is also a subsort of `Marble`. Magic potential is represented abstractly by a sort `Potential`. Four different potentials are defined (the second `ops` declaration):

- `+`, `-` represent positive and negative potentials, respectively
- `*`, `@` represent the potential of an activator marble to generate a positive or negative potential, respectively

A marble with potential p is constructed by annotating a plain marble m with p, written `[m | p]`.

```
fmod MAGIC-MARBLES-DATA is
  sort Marble .
  sort PlainMarble . subsort PlainMarble < Marble .
  ops redM blueM greenM purpleM whiteM blackM
       : -> PlainMarble [ctor] .

  sort MagicMarble .  subsort MagicMarble < Marble .

  sort Potential .
  ops + - * @ : -> Potential [ctor] .
  op `[_|_`] : PlainMarble Potential -> MagicMarble [ctor] .
endfm
```

The declaration beginning `op '[_|_']` is an example of *mixfix* syntax, where an operator is a collection of symbols including blanks (_) that are placeholders for the arguments. In the preceding case, there are two arguments—the first of sort `PlainMarble`, the second of sort `Potential`. (The backquotes are included to ensure the brackets are parsed as part of the operator symbol.) As examples, we have

- `[whiteM | *]` A white marble with positive activator potential
- `[whiteM | @]` A white marble with negative activator potential
- `[greenM | +]` A green marble with positive potential

The module `MAGIC-MARBLES-STATE` extends `MAGIC-MARBLES-DATA` (using the inclusion statement beginning `inc`) specifying a sort `Mix` of multisets of marbles. (A multiset or bag is a collection of elements where the number of occurrences of a given element matters, but the order does not.) The constant `none` represents the empty multiset, and multiset union is represented by the binary operator `_ _`, which is associative and commutative with identity `none` (ACI).

```
fmod MAGIC-MARBLES-STATE is
  inc MAGIC-MARBLES-DATA .
sort Mix .
subsort Marble < Mix .
op none : -> Mix [ctor] .
op _ _ : Mix Mix -> Mix [assoc comm id: none] .
endfm
```

Thus `redM blueM [whiteM | *]` is a mix of three marbles, two plain and one with positive activating potential.

The syntax `_ _` is an example of empty syntax—that is, a special case of mixfix syntax in which application of the operator relies on juxtaposition. The attribute `[assoc comm id: none]` at the end of the declaration specifies the ACI property, which axiomatizes the notion of multiset. Associative means that the two ways of joining three elements with two applications of the operator give the same multiset. For example:

```
(redM blueM) whiteM = redM (blueM whiteM) .
```

In this case, grouping of arguments doesn't matter, so the parentheses can be omitted. Commutative means that permuting the order of arguments gives the same result. For example:

```
redM blueM = blueM redM .
```

The "identity `none`" property means that adding `none` to the mix does not change the mix. For example:

```
redM blueM none = blueM redM .
```

The module `MAGIC-MARBLES-RULES` specifies how marbles interact using three rules. A rule begins with the keyword `rl` followed by the rule label enclosed in `[]` (brackets). The left-hand side (premise) and right-hand side (conclusion) of a rule are separated by the `=>` sign. The rules labeled `plus` and `minus` formalize the informal statements "A positive (resp., negative) activator marble can give positive (resp., negative) potential to a plain marble," "A marble with the * potential is a positive activator," and "A marble with the @ potential is a negative activator." The rule labeled `cancel` formalizes the statement "If a marble with negative potential contacts a marble with positive potential, the potential is cancelled and they both become plain."

```
mod MAGIC-MARBLES-RULES is
  inc MAGIC-MARBLES-STATE .
  vars pm0 pm1 : PlainMarble .
  rl[plus]:  pm0 [ pm1 | *] => [ pm0 | + ] [ pm1 | @ ] .
  rl[minus]:  pm0 [ pm1 | @] => [ pm0 | - ] [ pm1 | * ] .
  rl[cancel]:  [pm0 | +] [ pm1 | -] => pm0  pm1 .
endm
```

The variables `pm0`, `pm1` stand for arbitrary plain marbles. The change of activator potential in rules `plus,minus` formalize the rule, "When it does so, it changes parity and becomes a negative (resp., positive) activator."

Now we have an executable formal specification of magical marbles. What can we do with it? The simplest thing is to pick a starting state and use the rewrite and continue commands to watch it run. The command `rew [1] t .` rewrites the term `t` in one step. The command `cont 1 .` continues rewriting one more step. Suppose we have an initial state `[whiteM | *] redM blueM` with two plain marbles and a positive activator. The rule `plus` applies to the subterm `[whiteM | *] redM`, matching `pm0` to `redM` and `pm1` to `whiteM` (using multiset matching where the order of multiset elements doesn't matter), and replacing the matched subterm by the corresponding instance of the rule's right-hand side, `[redM | +] [whiteM | @]`.

```
Maude> rew [1] [whiteM | *] redM blueM .
result Mix: blueM [redM | +] [whiteM | @]
```

Rewriting can be continued by rewriting with the `minus` rule and then with the `cancel` rule.

```
Maude> cont 1 .
result Mix: [redM | +] [blueM | -] [whiteM | *]    *** by [minus]
Maude> cont 1 .
result Mix: redM blueM [whiteM | *]                *** by [cancel]
```

This computation could be continued for as many steps as you like. Many other computations are possible starting from our initial state, each making different choices of which plain marble to use in the `plus` step.

The command `search [n] istate =>+ pattern` searches the states reachable from `istate` for states matching `pattern`, stopping either when it has found `n` solutions or when it runs out of states. It starts by finding all states that result from application of one rewrite rule, then finding all states that result from application of one rewrite rule to each of these states, and so on. For example, using the search command, we can ask whether it is possible to make the red and blue marbles simultaneously positive, starting from our initial state.

```
Maude> search [1] [whiteM | *] redM blueM
       =>+ M:Mix [redM | + ] [blueM | + ] .
Maude> no Solution .
```

The answer is no. If we add another positive activation, then getting two positive marbles is easy, as we will see in a later example.

An alternative to searching is to use model checking. A model checker checks properties of the possible computations starting from a given initial state. The properties are expressed in Linear Temporal Logic (LTL). The module `MAGIC-MARBLES-MC` defines model checking states for Magic Marbles and defines two state propositions. The `{_}` operator encapsulates a mix, thereby defining a boundary. Propositions are defined using the relation, {mx}|= `prop`, read as "the mix `mx` satisfies the proposition `prop`." A predicate on mixes (and other arguments) can easily be turned into a proposition, by defining a corresponding operator that maps the remaining arguments to the sort `Prop`.

As an example, we define the function `contains` using two equations.

```
var mx mx0 mx1 : Mix .
op contains : Mix Mix -> Bool .
eq contains(mx0 mx1, mx0) = true .
eq contains(mx,mx0) = false [owise] .
```

The first equation uses multiset matching to identify the true case. The pattern `mx0 mx1` matches a mix term `mx` only if `mx` contains `mx0` (`mx1` is the rest of the mix). In this case `contains(mx,mx0)` will be rewritten as `true`. An equation, such as the second equation given previously, when tagged with the `[owise]` attribute, can be used only if no other equation applies. In our case, the second equation applies to a term `contains(mx,mx0)` only if `mx` does not contain `mx0`. In this case, the term is rewritten as `false`. Then `hasMix(mx0)`, a proposition corresponding to the predicate `contains(mx,mx0)`, is defined such that {mx} satisfies `hasMix(mx0)` if `contains(mx,mx0)`. We also define a proposition `nPot(n,p)` such that {mx} satisfies `nPot(n,p)` if `count(mx,p) == n`, where `count(mx,p)` is the number `n` of marbles in `mx` with potential `p`.

```
mod MAGIC-MARBLES-MC is
  inc MAGIC-MARBLES-RULES .
```

```
  inc MODEL-CHECKER .

  op '{_'} : Mix -> State .

  vars mx mx0 : Mix .   var p : Potential .   var n : Nat .
  op nPot : Nat Potential -> Prop .
  eq {mx} |= nPot(n,p) = count(mx,p) == n .
  op hasMix : Mix -> Prop .
  eq {mx} |= hasMix(mx0) = contains(mx,mx0) .
endm
```

If `P` is a state proposition, then the property `[]P` says that every state in a computation satisfies `P` and `[]~P` says that no state in a computation satisfies `P`. Thus, to see if a state satisfying `P` can be reached, we can use the Maude model checker to evaluate **`modelCheck({mx},[]~P)`**. If a state can be reached satisfying `P`, the model checker will return a counterexample showing the transitions (state and rule label) of a computation containing such a state. For example, starting from a state with two plain red marbles, two plain blue marbles, and two positive activator marbles, we can ask the following questions: (1) Can we get one red and one blue marble with positive potential? (2) Can we get three marbles with positive potential? and (3) What about four? The following is the model-checking query to answer the first question.

```
Maude> red modelCheck(
  {[whiteM | *] [blackM | *] redM redM blueM blueM},
                        [] ~ hasMix([redM | +] [blueM | +]) ) .
reasult ModelCheckResult: counterexample(
{{redM redM blueM blueM [whiteM | *] [blackM | *]},'plus}
{{redM blueM blueM [redM | +] [whiteM | @] [blackM | *]},'plus}
{{blueM blueM [redM | +] [redM | +] [whiteM | @] [blackM | @]},
                                                          'minus}
{{blueM [redM | +] [redM | +] [blueM | -]
                           [whiteM | *] [blackM |  @]},'plus}
{{[redM | +] [redM | +] [blueM | +] [blueM | -]
                  [whiteM | @] [blackM | @]},'cancel}, ....)
```

A counterexample is a list of transitions that fails to satisfy the property being model checked. A transition is represented by the starting state and the label of the rule applied. Thus the starting state of the first transition in the preceding counterexample is the initial state, and the **`plus`** rule is applied to one of the red marbles. The second transition applies the **`plus`** rule to the other red marble. This is followed by the **`minus`** transition to recover a positive activator marble, and finally the remaining blue marble is made postive. The ellipses (...) indicate that the counterexample continues. This is an artifact of the model-checker requirement that counterexamples are infinite. The remainder of the computation entails building a loop and can be ignored for our purposes.

The second and third questions are answered by the following queries. In fact, the counterexample for the first question also answers the second question. The true result for the third query says that, in fact, it is not possible to obtain four marbles with positive potential given our initial state.

```
Maude> red modelCheck(
    {[whiteM | *] [blackM | *] redM redM blueM blueM},
     [] ~ nPot(3,+)) .
result:  same as for (1)

Maude> red modelCheck(
    {[whiteM | *] [blackM | *] redM redM blueM blueM},
     [] ~ nPot(4,+)) .
result Bool: true
```

1.4 Building a Pathway Logic Knowledge Base

Now we describe a small PL knowledge base, SmallKB, that represents initial signaling events in response to epidermal growth factor (Egf) stimulation. Egf initiates epidermal growth factor receptor (EgfR) signaling, which regulates growth, survival, proliferation, and differentiation in mammalian cells. In particular, we will look at an initial part of the mitogen-activated protein kinase (MAPK) pathway [4, 12, 24, 45].[1] This pathway is often represented by the following linear sequence:

$$\text{Egf} \rightarrow \text{EgfR} \rightarrow \text{Grb2} \rightarrow \text{Sos1} \rightarrow \text{Ras} \rightarrow \text{Raf1} \rightarrow \text{Mek} \rightarrow \text{Erk}$$

Recall that a rewriting logic specification has two parts: an equational part specifying the structure and static properties of system states, and a rules part specifying the system behaviors. A PL system is structured in four layers: (1) sorts and operations, (2) molecular components, (3) rules, and (4) queries. Layers 1 through 3 make up a pathway knowledge base (PL KB), with layers 1 and 2 being the equational part. Layer 4 specifies models (initial states of interest).

1.4.1 The Equational Part

The *sorts and operations* layer declares the main sorts, subsort relations, and operators to construct representations of cellular states. The sorts of entities include Chemical, Protein, Complex, Location (position within cellular compartments), and Cell. These are all subsorts of the sort Soup, which represents "liquid" mixtures as multisets of entities. The sort Modification is used to represent post-translational protein modifications.

---

[1] The Wikipedia article on signal transduction (found at `http://en.wikipedia.org/wiki/Signal_transduction`) contains an excellent overview and is a good place for the reader who is not familiar with cellular signaling notions to start reading to learn more [23].

They can be abstract, to specify that a protein is activated, bound, or phosphorylated, or more specific, such as phosphorylation at a particular site. Modifications are applied using the operator `[_-_]`. (Note the similarity to the annotation of marbles with potential in Section 1.3.) For example, the term `[Raf1 - act]` represents `Raf1` in an activated state, and `[Hras - GTP]` represents the protein `Hras` in its "on" state (loaded with GTP). (Hras is a specific member of the Ras family.) The term `[Gab1 - Yphos]` represents `Gab1` phosphorylated on a tyrosine site, while `[Gab1 - phos(Y 627)]` represents `Gab1` phosphorylated on tyrosine 627. Complex formation is represented by the operation `(_:_)`. For example, the term `(Egf : [EgfR - act])` represents the complex resulting from binding of Egf to EgfR and subsequent activation of EgfR. A cell state is represented by a term of the form

```
[cellType | locs] .
```

The symbol `cellType` specifies the type of cell—for example, a macrophage or fibroblast. The symbol `Cell` is used to indicate an unspecified cell type. The symbol `locs` represents the contents of a cell organized by cellular location. Each location is represented by a term of the form `{ locName | components }`, where `locName` identifies the location—for example, `CLm` stands for cell membrane, and `components` stands for the mixture of proteins and other compounds in that location. For example:

```
[Cell | {CLm | EgfR  PIP2}
        {CLi | [Hras - GDP]  Src}
        {CLc | Gab1 Grb2 Pi3k Plcg Sos1}]) .
```

represents a generic cell with three locations: the membrane (location tag `CLm`) contains `EgfR` and a chemical `PIP2` (discussed later in this section); the inside of the membrane (location tag `CLi`) contains `Hras` loaded with `GDP` and `Src`; and the cytoplasm (location tag `CLc`) contains `Gab1`, `Grb2`, `Pi3k`, `Plcg`, and `Sos1`.

The *components* layer specifies particular entities (proteins, genes, chemicals) and introduces additional sorts for grouping proteins by family or shared behavior. For example, `ErbB1L` is a subsort of `Protein` whose elements are ErbB1 (`EgfR`) ligands. Components are declared as constants, giving their sort, and their metadata provide synonyms and standard names that can be linked to databases providing other information. For example, the epidermal growth factor `Egf` with sort `ErbB1L`, and metadata giving its HUGO and Swiss-Prot names, its *category*, and two synonyms, is declared as follows:

```
op Egf : -> ErbB1L [metadata "(\
  (spname EGF_HUMAN)\
  (spnumber P01133)\
  (hugosym EGF)\
```

```
  (category Ligand)\
  (synonyms \"Pro-epidermal growth factor precursor, EGF\" \
       \"Contains: Epidermal growth factor, Urogastrone \"))"] .
```

Similarly, `EgfR` is declared simply to be a protein:

```
op EgfR : -> Protein [metadata "(\
  (spname EGFR_HUMAN)\
  (spnumber P00533)\
  (hugosym EGFR)\
  (category Receptor)\
  (synonyms \"Epidermal growth factor receptor precursor\" \
       \"Receptor tyrosine-protein kinase ErbB-1, ERBB1 \"))"] .
```

PIP2 is a chemical (a lipid) residing in the membrane. Its phosphorylated form, PIP3, plays an important role in a number of signaling pathways, either directly or through its cleavage products. Chemicals have metadata linking them to a KEGG database entry, where much information can be found.

```
op PIP2 : -> Chemical [metadata "(\
   (category Chemical)\
   (keggcpd C04569)\
   (synonyms \"Phosphatidylinositol-4,5P \" ))"] .
```

The *rules* layer is the heart of a PL KB. It contains rewrite rules specifying individual reaction steps. In the case of signal transduction, these rules represent processes such as activation, phosphorylation, complex formation, and translocation. The rules layer is discussed in Section 1.4.2.

The *queries* layer specifies initial states (called dishes) to be studied. Initial states can be thought of as describing "in silico experiments." They represent in silico Petri dishes containing a cell and ligands of interest in the surrounding mixture. An initial state is represented by a term of the form

```
PD(out cell)
```

where `cell` represents a cell state and `out` represents a soup of ligands and other molecular components in the cell's surroundings. Although a dish can contain many cells, the current PL analysis tools can handle only single cells. For example, an initial state to study Ras activation in `SmallKB` is given by the dish term

```
op rasDish  : -> Dish .
eq rasDish =
     PD(Egf [Cell | {CLm | EgfR  PIP2}
                    {CLi | [Hras - GDP]  Src}
                    {CLc | Gab1 Grb2 Pi3k Plcg Sos1}]) .
```

representing a dish containing `Egf` and the cell discussed previously.

1.4.2 The Rules Part

PL rules are curated from the literature, and each rule has associated evidence items describing experimental data that serve as evidence for the rule. Discussion of evidence is beyond the scope of this chapter, as it requires some understanding of experimental methods to be meaningful. Rule 1 (label `1.EgfR.act`) describes the binding of an ErbB1 ligand to EgfR, formalizing the first step of the signaling sequence given earlier.

```
rl[1.EgfR.act]:
  ?ErbB1L:ErbB1L
  [?CellType:CellType | ct {CLm | clm EgfR}]
  =>
  [?CellType:CellType | ct
    {CLm | clm ([EgfR - act] : ?ErbB1L:ErbB1L)} ] .
```

The term `?ErbB1L:ErbB1L` is a variable that matches any ErbB1 ligand—for example, `Egf`—and `?CellType:CellType` is a variable that matches any cell type. `{CLm | clm EgfR}` matches any cell membrane location that contains `EgfR`, because `clm` is a variable that will match the rest of the membrane contents. Thus the left-hand side subterm

```
[?CellType:CellType | ct CLm | clm EgfR]
```

matches any cell that contains `EgfR` in the cell membrane, because the variable `ct` will match any additional locations. For example, it matches the initial state `rasDish` with

```
?CellType:CellType := Cell
?ErbB1L:ErbB1L := Egf
clm := PIP2
ct := {CLi | [Hras - GDP]  Src} {CLc | Gab1 Grb2 Pi3k Plcg Sos1}
```

The result of rewriting `rasDish` with rule 1 is

```
PD([Cell |
   {CLm | ([EgfR - act] : Egf)  PIP2}
   {CLi | [Hras - GDP]  Src}
   {CLc | Gab1 Grb2 Pi3k Plcg Sos1}]) .
```

which is obtained by instantiating the rule's right-hand side using the variable bindings from the left-hand side match.

Once the receptor is activated, it can recruit `Grb2` to the membrane interior. This situation is describe by the rule labeled `5.Grb2.reloc`.

```
rl[5.Grb2.reloc]:
  {CLm | clm ([EgfR - act] : ?ErbB1L:ErbB1L)  }
  {CLi | cli                          }
  {CLc | clc Grb2                     }
  =>
```

```
  {CLm | clm ([EgfR - act] : ?ErbB1L:ErbB1L) }
  {CLi | cli [Grb2 - reloc]    }
  {CLc | clc                   } .
```

Notice that on the left, `Grb2` is in the `CLc` location (cytoplasm); on the right, it is in the `CLi` location. The modification `reloc` makes the change in location explicit. It is not strictly necessary, but makes the changes easier to follow. Continuing the rewriting of `rasDish` with rule `5.Grb2.reloc`, we get

```
PD([Cell |
   {CLm | ([EgfR - act] : Egf)  PIP2}
   {CLi | [Hras - GDP]  Src [Grb2 - reloc]}
   {CLc | Gab1 Pi3k Plcg Sos1}]) .
```

Now `Sos1` can be recruited to the membrane complex by binding to `Grb2`. This is described by rule `13.Sos1.reloc`. Note that in this particular representation, the complex formation is abstracted to colocation. We could also make the complex explicit if that understanding were needed for some analysis.

```
rl[13.Sos1.reloc]:
  {CLi | cli [Grb2 - reloc]                }
  {CLc | clc Sos1                          }
  =>
  {CLi | cli [Grb2 - reloc] [Sos1 - reloc] }
  {CLc | clc                               } .
```

The resulting state is

```
PD([Cell |
   {CLm | ([EgfR - act] : Egf)  PIP2}
   {CLi | [Hras - GDP]  Src [Grb2 - reloc] [Sos1 - reloc]}
   {CLc | Gab1  Pi3k Plcg }]) .
```

In the next section we will see how to turn the Maude terms into a graph representation that makes it easier to visualize and understand reaction networks and their evolution. In particular, a graphical representation of the preceding three-step computation is shown in Figure 1.2.

1.5 Computing with a PL KB

The Pathway Logic Assistant (PLA) provides interactive graphical access to a PL knowledge base. For this purpose, rule sets and computations are represented using Petri nets [37,38,46], which have a natural graphical representation. Additionally, very efficient tools are available for analyzing the Petri net models generated by PLA.

1.5.1 PL Petri Nets

Petri nets were invented to model execution of concurrent processes and, therefore, are nicely suited to modeling signals propagating through a cell. A Pathway Logic Petri net (or simply, a Petri net) can be thought of as a graph with two kinds of nodes: rule nodes (shown as squares) and occurrence nodes (shown as ovals). Rule nodes, called transitions in the Petri net community, represent reactions; occurrence nodes, called places in the Petri net community, represent reactants, products, or modifiers. Occurrences can be thought of as atomic propositions asserting that a protein (in a given state) or other component (small molecule, complex, etc.) occurs in a given compartment. In this view, rules are logical implications.

An occurrence oval is labeled by a string representation of the corresponding Maude term. For example, the string representation of `Efg` outside a cell is `Egf-Out` and the string representation of `Egf : [EgfR - act]` in the cell membrane is `Egf:EgfR-act-CLm`. The reactants of a rule are the occurrences connected to the rule by arrows from the occurrence to the rule. The products of a rule are the occurrences connected to the rule by arrows from the rule to the occurrence. The modifiers of a rule (enzymes and other components that must be present but are unchanged) are the occurrences connected to the rule by a dashed arrow. As an example, the Petri net representation of the rule for recruitment of `Sos1` is shown in Figure 1.1.

The rule is represented by the rectangle labeled `13` (the short form of `13.Sos1.reloc`). The reactant `Sos1` in the cytoplasm is represented by the oval labeled `Sos1-CLc` with an arrow from the oval to the rule rectangle. The product `[Sos1 - reloc]` at the membrane interior is represented by the oval labeled `Sos1-reloc-CLi` with an arrow from the rule rectangle to the oval. `[Grb2-reloc]` drives the reaction but is not changed (at our level of representation); thus it is represented by the oval labeled `Grb2-reloc-CLi` with a dashed arrow from the oval to the rule rectangle.

A set of Petri net rules corresponding to the rules of a PL knowledge base is called a transition knowledge base (TKB). The analog of a PL dish is a PL Petri net state, which specifies the occurrences that are present; that

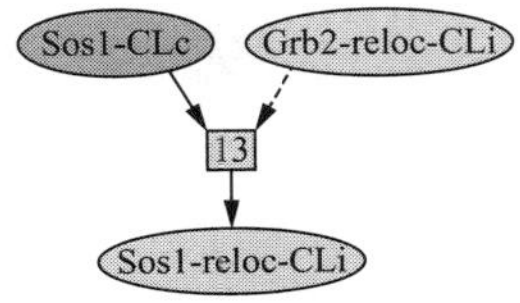

Figure 1.1 Petri net transitions for rule 13.Sos1.reloc

is, it specifies the state and location of each molecular component. Given a state, a Petri net rule is enabled if all of its occurrences connected by incoming arrows (reactants and modifiers) are present in the state. When an enabled rule fires, the reactant occurrences are removed from the state and the product occurrences are added. The modifier occurrences are left unchanged.

Corresponding to a PL model, a Petri net model consists of a set of rules (a TKB) and an initial state. To execute a Petri net model, one puts tokens on the ovals corresponding to occurrences present in the initial state, and moves tokens as rules become enabled and fire. Figure 1.2 shows the execution of a Petri net model of the process that recruits `Sos1` to the membrane interior.

There are three rules (corresponding to the rewrite rules discussed in Section 1.4). Darker ovals represent occurrences that are present (marked with a token). Figure 1.2(a) shows the initial state, with `EgfR-CLm`, `Egf-Out`, `Grb2-CLc`, and `Sos1-CLc` marked as initially present. The only rule enabled is rule 1. Figure 1.2(b) shows the result of firing rule 1, which removes tokens from `EgfR-CLm` and `Egf-Out` and adds a token to `Egf:EgfR-act-CLm`. Now rule 5 is enabled and Figure 1.2(c) shows the result of firing rule 5, which enables rule 13. Figure 1.2(d) shows the final state.

Starting with a PL KB, we convert it to a Petri net TKB, and convert dishes to occurrence sets, in a way that preserves the possible executions. We can then analyze models by finding subnets relevant to a desired state (goal), finding pathways reaching a goal, comparing subnets and/or pathways, finding knockouts (omissions from the initial state that prevent reaching a goal), or exploring a network of rules by incrementally adding connected components and rules to a given starting subnet. This process is explained briefly in the following subsections. Details, including proof that the Petri net representation is equivalent to the rewriting logic representation, can be found in [49].

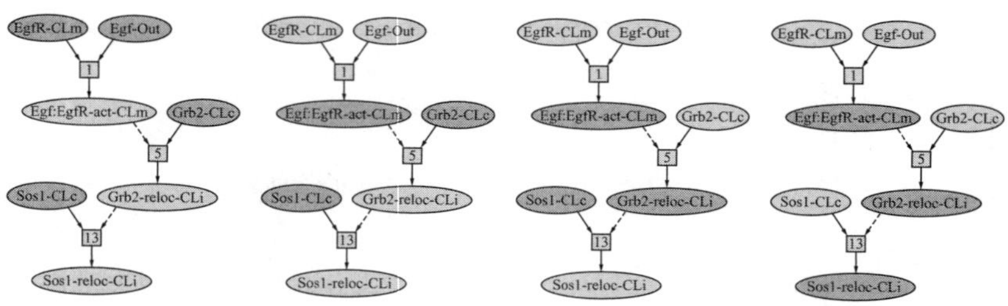

(a) Initial state (b) Step 1 (c) Step 2 (d) Step 3

Figure 1.2 Execution of the Sos1 recruitment pathway

1.5.2 Converting a PL KB to a Petri Net TKB

Transformation to Petri net representation accomplishes several things. First, it supports graphical representation of the data set. Second, it makes things concrete. The full PL representation allows for rules that express families of reactions and for multiple cells and cell types. In PLA, we restrict attention to systems with a single cell. For each variable that stands for a single component, we fix a specific (finite) set of components that can be values of that variable. For example, in the SmallKB knowledge base, two proteins can be values of the variable ?ErbB1L:ErbB1L of sort ErbB1L—namely, Egf and Tgfa. Producing a Petri net representation of the PL knowledge base proceeds in two steps. The first step transforms rules into occurrence form, by transforming the dishes or cells appearing in PL rules into occurrence sets. The second step instantiates remaining variables with known values.

Formally, an occurrence is a pair consisting of a component (a protein, possibly modified; a small molecule; or a complex) and a location name. For example, <Egf, Out> is an occurrence representing Efg outside a cell and < Egf : [EgfR - act], CLm> is an occurrence representing Egfr complexed with Egf and activated in the cell membrane. The left or right side of a rule is transformed by pairing each component with its location (the name of the enclosing location), dropping the location container, and dropping variables such as ct or clm that serve only to name location contents that are not important for the rule. Thus rule 1 (1.EgfR.act) becomes

```
rl[1.EgfR.act.pn]
    < ?ErbB1L:ErbB1L, Out >  < EgfR , CLm >
   =>
   < ?ErbB1L:ErbB1L  : [EgfR - act], CLm >
```

When we instantiate the remaining variables, we also convert rules (logical statements) into elements of a data type called PNetTransition. This allows us to compute with and reason about the Petri net rules directly. An interpreter to execute rules represented as data can be defined by a single rewrite rule. A Petri net transition term has the form

```
pnTrans(label,iOccs,oOccs,bOccs)
```

where label is a quoted identifier, and iOccs, oOccs, and bOccs are multisets of occurrences: iOccs are the occurrences required and removed by the transition (connected to the rule by incoming arrows), oOccs are the occurrences produced by the transition (connected to the rule by outgoing arrows), and bOccs are the occurrences required but not removed by the transition (connected to the rule by dashed arrows). As an example, two Petri net transitions are obtained by instantiating the occurrence form of rule 1. The first is obtained by instantiating the variable `?ErbB1L:ErbB1L` with `Egf`:

```
pnTrans('1.EgfR.act,
        < Egf,Out > < EgfR,CLm >,
        < Egf :[EgfR -  act],CLm >,
        none)
```

The second is obtained by instantiating with `Tgfa`:

```
pnTrans('1.EgfR.act#1,
       < EgfR,CLm > < Tgfa,Out >,
       < Tgfa :[EgfR - act],CLm >,
       none)
```

The transition label is the rule label, suffixed with `#1`, `#2`, ... if there are multiple instantiations. In rule 1 there are no unchanged occurrences, so `iOccs` is simply the instantiated occurrences from the rule's left-hand side, `oOccs` is the instantiated occurrences from the rule's right-hand side, and `bOccs` is `none`, the empty occurrence set.

As another example, the Sos1 recruitment rule (`13.Sos1.reloc`) is transformed into the following Petri net transition:

```
pnTrans('13.Sos1.reloc,
       < Sos1,CLc >,
       <[Sos1 - reloc],CLi >,
       <[Grb2 - reloc],CLi >)
```

In this case, `bOccs` is `<[Grb2 - reloc],CLi >`, which is necessary for the rule to fire, but not be used up.

The process of converting a rule set into a list of Petri net transitions uses Maude's meta-level, where rules are represented as data and one can manipulate terms with variables (which are also just data in the meta-level).

1.5.3 PL Petri Net Models

Once we have a TKB, we can derive models and perform computations with them—for example, asking for subnets, pathways, and knockouts, and making comparisons. A model consists of a Petri net transition list (specifying the possible transitions) and a set of occurrences representing the initial (or current) state. It is derived from a dish and a TKB by transforming the dish into a set of occurrences and doing a *forward collection* in the TKB from the occurrence set to derive the set of transitions that could possibly be enabled in a computation starting from the initial state. The idea of the forward collection is to iteratively augment the occurrence set with all occurrences that could be produced by firing enabled transitions (from the TKB), without removing the `iOccs` part of the transition, and then add enabled transitions to the accumulating list of transitions.

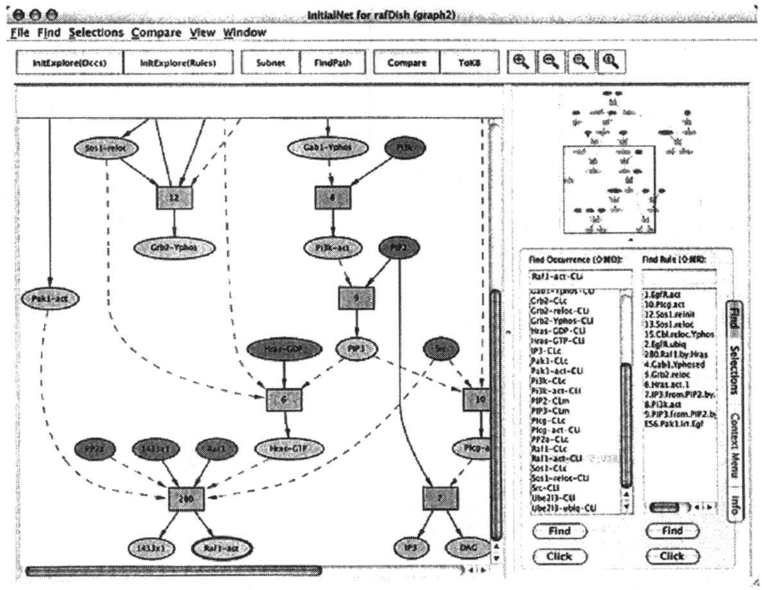

Figure 1.3 Raf activation model viewed in PLA

The result of forward collection is a transition list that is possibly an over-approximation. That is, any transition of TKB that becomes enabled in some computation from the initial state will be in the accumulated transition list, but some transitions may not become enabled in any computation from the initial state. The crucial point is that we don't lose any possible computations by restricting the set of transitions to be considered, and the simple over-approximation makes the model derivation feasible.

Figure 1.3 shows a screenshot of the Petri net model of Raf activation, generated by PLA from the dish `rafDish` whose occurrence set is `rafDishOccs`:

```
eq rafDishOccs =
  < Egf, Out > < EgfR, CLm >  < PIP2, CLm >
  < [Hras - GDP], CLi> < Src, CLi> < [Ube213 - ubiq], CLi >
  < Cbl, CLc >  < Gab1, CLc >  < Grb2, CLc >  < Pi3k, CLc >
  < Plcg, CLc >  < Sos1, CLc >  < 1433x1, CLc >  < Pak1, CLc >
  < Raf1, CLc >  < PP2a, CLc > .
```

As discussed earlier, ovals are occurrences, with initial occurrences having darker shading. Rectangles are transitions. Dashed arrows indicate an occurrence that is both input and output. The thumbnail sketch in the upper-right corner of the figure shows the full network. The main frame shows a magnified version of the portion of the network in the red rectangle. To change the view in the main frame, the red rectangle can be dragged around in the thumbnail frame. The view can also be changed using the scroll bars. The Finder in the lower-right corner allows the user to locate occurrences and rules by name, and to center the view on the selected node.

PLA provids a simple query language for specifying signaling pathways of interest. A query specifies three sets: goals, avoids, and hides. Goals are a set of occurrences that should appear at the end of a pathway; they represent properties of a desired state. Avoids are a set of occurrences that should not appear in any state in the execution of the pathway. Hides are a set of rules that should not fire in the pathway. To create a query, goals, avoids, and hides sets can be selected by clicking the desired occurrence or rule and then pressing the corresponding button in the information window that appears. Once query elements have been selected, the user can ask to see the relevant subnet or to find a path. The relevant subnet contains all of the rules needed for any (minimal) pathway satisfying the query, while the path is just the first path found by the analysis tool. The relevant subnet is computed directly from the Petri net transitions list, together with the initial state and query elements in a manner similar to the forward collection function described earlier.

The logic underlying the query language involves a temporal logic. Goal queries are answered by model checking the assertion that the goal set is not reachable from the initial state `ioccs` in a transition list `pntl*` from which transitions that produce an avoid or are in the hides set are removed.

```
(pntl*,ioccs) |= []~ goal
```

A pathway satisfying a query is obtained by translating the reduced Petri net transition list and query into the language of the Low-Level Analyzer (LoLA) model checker [29, 44], asserting that no such pathway exists. If a pathway does exist, LoLA returns a list of transitions in the pathway, which

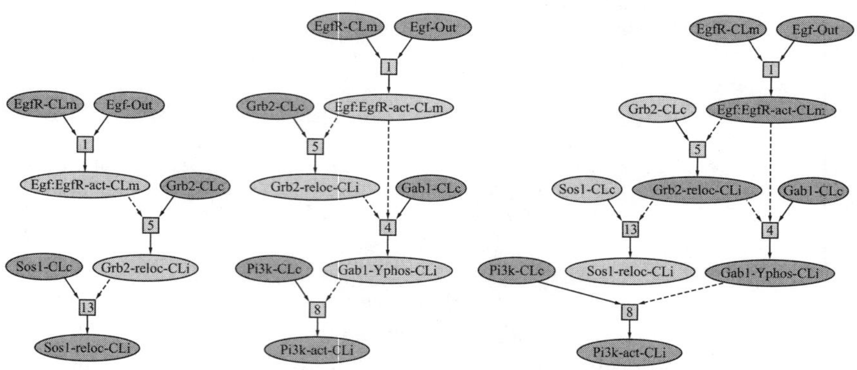

(a) Sos1 path (b) Pi3k path (c) Comparison

Figure 1.4 Sos1 and Pi3k activation paths and comparison

PLA converts to a Petri net for display and possible further analysis. The LoLA model checker is highly optimized for Petri nets and, therefore, can be used to perform computations with very large models.

Figure 1.4 shows pathways in the Raf1 model that recruit `Sos1` (a) and that activate `Pi3k` (b), which are obtained by making `Sos1-reloc-CLi` or `Pi3k-act-CLi` a goal (indicated by coloring the oval green) and using Find-Path. The key property of a pathway is that executing the pathway Petri net starting from the initial state leads to a state in which the goals are found among the occurrences—that is, a state satisfying the goal is reached. Furthermore, none of the avoids or hides appears in the pathway.

In addition to generating pathway subnetworks, two subnets can be compared. To do so the two networks are merged into one. Figure 1.4(c) shows the result of comparing the `Sos1` and `Pi3k` pathways. Nodes in both pathways are colored pink, nodes only in the `Sos1` pathway are colored cyan, and nodes only in the `Pi3k` pathway are colored dark lavender.

The `Sos1` and `Pi3k` pathways are part of the model of Hras activation, and ultimately the model of Raf activation. Figure 1.5(a) shows a pathway activating Hras, obtained by specifying `Hras-GTP-CLi` as a goal. Figure 1.5(b) shows the `Sos1` and `Pi3k` comparison as a subnet of the `Hras` activation pathway (nodes only in the `Hras` path are white).

The SmallKB and the Ras and Raf1 activation initial states are available as part of the Pathway Logic Demo found on the Pathway Logic website (`http://pl.csl.sri.com/`), along with papers, tutorial material, and download of the Pathway Logic Assistant tool.

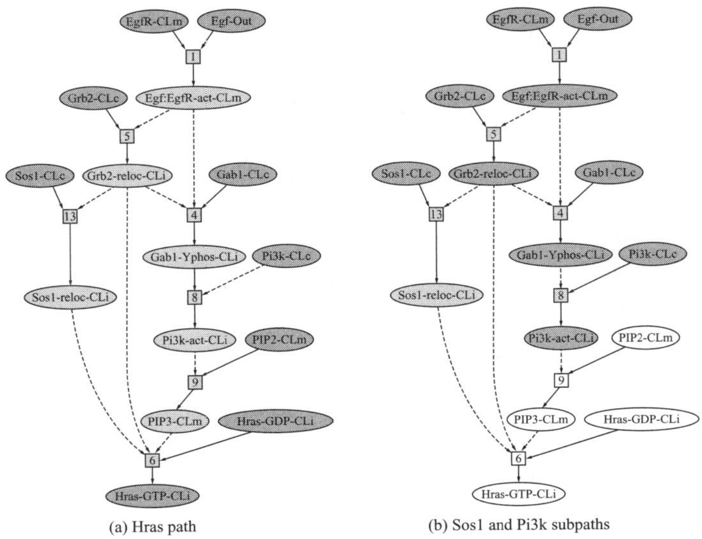

Figure 1.5 Hras activation pathway and Sos1, Pi3k subnets

1.6 Conclusion

Pathway Logic is a symbolic systems biology approach to modeling biological processes based on rewriting logic. This chapter described the use of Pathway Logic to model signal transduction processes, and the use of the Pathway Logic Assistant to browse and analyze these models. Pathway Logic can also be used to model and analyze metabolic networks and to interpret experimental data. Future challenges include integration of signaling and metabolic network models, and new abstractions to simplify networks and identify meaningful modules.

References

[1] M. Calder, V. Vyshemirsky, D. Gilbert, and R. Orton. Analysis of signalling pathways using the PRISM model checker. In G. Plotkin, editor, *Proceedings of the Third International Conference on Computational Methods in System Biology*, 2005.

[2] L. Calzone, N. Chabrier-Rivier, F. Fages, L. Gentils, and S. Soliman. Machine learning bio-molecular interactions from temporal logic properties. In G. Plotkin, editor, *Proceedings of the Third International Conference on Computational Methods in System Biology*, 2005.

[3] L. Cardelli. Abstract machines of systems biology. In *Transactions on Computational Systems Biology III*, volume 3737 of *Lecture Notes in Computer Science*, pages 145–168, 2005.

[4] L. Cardelli. Brane calculi interactions of biological membranes. In *Computational Methods in Systems Biology*, volume 3082 of *Lecture Notes in Computer Science*. Springer, 2004.

[5] N. Chabrier-Rivier, M. Chiaverini, V. Danos, F. Fages, and V. Schächter. Modeling and querying biomolecular interaction networks. *Theoretical Computer Science*, 351(1):24–44, 2004.

[6] C. Chaouiya. Petri net modelling of biological networks. *Briefings in Bioinformatics*, 8:210–219, 2007.

[7] M. Clavel, F. Durán, S. Eker, P. Lincoln, N. Mart´-Oliet, J. Meseguer, and C. Talcott. *All About Maude: A High-Performance Logical Framework*. Springer, 2007.

[8] J. S. Edwards, M. Covert, and B. O. Palsson. Metabolic modelling of microbes: The flux-balance approach. *Environmental Microbiology*, 4(3): 133–140, 2002.

[9] S. Efroni, D. Harel, and I. R. Cohen. Towards rigorous comprehension of biological complexity: Modeling, execution and visualization of thymic t-cell maturation. *Genome Research*, 2003. Special issue on Systems Biology, in press.

[10] S. Eker, M. Knapp, K. Laderoute, P. Lincoln, J. Meseguer, and K. Sonmez. Pathway logic: Symbolic analysis of biological signaling. In *Proceedings of the Pacific Symposium on Biocomputing*, pages 400–412, January 2002.

[11] S. Eker, M. Knapp, K. Laderoute, P. Lincoln, and C. Talcott. Pathway logic: Executable models of biological networks. In *Fourth International Workshop on Rewriting Logic and Its Applications*, volume 71 of *Electronic Notes in Theoretical Computer Science*. Elsevier, 2002.

[12] F. Fages, S. Soliman, and N. Chabrier-Rivier. Modelling and querying interaction networks in the biochemical abstract machine BIOCHAM. *Journal of Biological Physics and Chemistry*, 4(2):64–73, 2004.

[13] J. Fisher and T. A. Henzinger. Executable cell biology. *Nature Biotechnology*, 25(11), 2007.

[14] H. Genrich, R. Küffner, and K. Voss. Executable Petri net models for the analysis of metabolic pathways. *Software Tools for Technology Transfer*, 3, 2001.

[15] R. Ghosh, A. Tiwari, and C. Tomlin. Automated symbolic reachability analysis with application to delta-notch signaling automata. In O. Maler and A. Pnueli, editors, *Hybrid Systems: Computation and Control HSCC*, volume 2623 of *Lecture Notes in Computer Science*, pages 233–248. Springer, April 2003.

[16] D. Gilbert, M. Heiner, and S. Lehrack. A unifying framework for modelling and analysing biochemical pathways using Petri nets. In M. Calder and S. Gilmore, editors, *CMSB*, volume 4695 of *Lecture Notes in Computer Science*, pages 200–216. Springer, 2007.

[17] P. J. Goss and J. Peccoud. Quantitative modeling of stochastic systems in molecular biology using stochastic Petri nets. *Proceedings of the National Academy of Science*, 95:6750–6755, 1998.

[18] D. Harel. Statecharts: A visual formalism for complex systems. *Science of Computer Programming*, 8:231–274, 1987.

[19] T.A. Henzinger. The theory of hybrid automata. In *11th IEEE Symposium on Logic in Computer Science*, pages 278–292, 1996.

[20] W. S. Hlavacek, J. R. Faeder, M. L. Blinov, R. G. Posner, M. Hucka, and W. Fontana. Rules for modeling signal-transduction systems. *Science STKE*, July 2006.

[21] R. Hofestädt. A Petri net application to model metabolic processes. *Systems Analysis Modelling Simulation*, 16:113–122, 1994.

[22] N. Kam, D. Harel, H. Kugler, R. Marelly, A. Pnueli, J. Hubbard, and M. Stern. Formal modeling of *C. elegans* development: A scenario-based approach. In *First International Workshop on Computational Methods in Systems Biology*, volume 2602 of *Lecture Notes in Computer Science*, pages 4–20. Springer, 2003.

[23] M. Knapp, personal communication, February 2008.

[24] W. Kolch. Meaningful relationships: The regulation of the Ras/Raf/MEK/ERK pathway by protein interactions. *Biochemical Journal*, 351:289–305, 2000.

[25] R. Küffner, R. Zimmer, and T. Lengauer. Pathway analysis in metabolic databases via differential metabolic display (DMD). *Bioinformatics*, 16:825–836, 2000.

[26] J. M. Kyriakis and J. Avruch. Mammalian mitogen-activated protein kinase signal transduction pathways activated by stress and inflammation. *Physiological Reviews*, 81:807–869, 2001.

[27] C. Li, Q. W. Ge, M. Nakata, H. Matsuno, and S. Miyano. Modelling and simulation of signal transductions in an apoptosis pathway by using timed Petri nets. *Journal of Bioscience*, 32:113–127, 2007.

[28] P. Lincoln and A. Tiwari. Symbolic systems biology: Hybrid modeling and analysis of biological networks. In R. Alur and G. Pappas, editors, *Hybrid Systems: Computation and Control HSCC*, volume 2993 of *Lecture Notes in Computer Science*, pages 660–672. Springer, March 2004.

[29] LoLA: Low-Level Petri net analyzer, 2004. `http://www.informatik.hu-berlin.de/~kschmidt/lola.html`.

[30] H. Matsuno, A. Doi, M. Nagasaki, and S. Miyano. Hybrid Petri net representation of gene regulatory network. *Pacific Symposium on Biocomputing*, 5:341–352, 2000.

[31] J. Meseguer. Conditional rewriting logic as a unified model of concurrency. *Theoretical Computer Science*, 96(1):73–155, 1992.

[32] R. Milner. *Communication and Concurrency*. Prentice-Hall, 1989.

[33] R. Milner. *Communicating and Mobile Systems: The Pi-Calculus*. Cambridge University Press, Cambridge, UK, 1999.

[34] F. Nielson, H. R. Nielson, C. Priami, and D. Rosa. Control flow analysis for bioambients. In *BioConcur*, 2003.

[35] Gh. Păun. *Membrane Computing: An Introduction*. Springer-Verlag, 2002.

[36] G. Pearson et al. Mitogen-activated protein (MAP) kinase pathways: Regulation and physiological functions. *Endocrine Reviews*, 153–183, 2001.

[37] J. L. Peterson. *Petri nets: Properties, Analysis, and Applications*. Prentice-Hall, 1981.

[38] C. A. Petri. Introduction to general net theory. In W. Brauer, editor, *Net Theory and Applications, Proceedings of the Advanced Course on General Net Theory of Processes and Systems, Hamburg, 1979*, volume 84 of *Lecture Notes in Computer Science*, pages 1–19. Springer-Verlag, 1980.

[39] C. Priami, A. Regev, E. Shapiro, and W. Silverman. Application of a stochastic name-passing calculus to representation and simulation of molecular processes. *Information Processing Letters*, 80:25–31, 2001.

[40] M. J. Pérez-Jiménez and F. J. Romero-Campero. Modelling EGFR signalling cascade using continuous membrane systems. In G. Plotkin, editor, *Proceedings of the Third International Conference on Computational Methods in System Biology*, 2005.

[41] V. N. Reddy, M. N. Liebmann, and M. L. Mavrovouniotis. Qualitative analysis of biochemical reaction systems. *Computational Biological Medicine*, 26:9–24, 1996.

[42] A. Regev, E. Panina, W. Silverman, L. Cardelli, and E. Shaprio. BioAmbients: An abstraction for biological compartments. *Theoretical Computer Science*, 325:141–167, 2004.

[43] A. Regev, W. Silverman, and E. Shapiro. Representation and simulation of biochemical processes using the pi-calculus process algebra. In *Pacific Symposium on Biocomputing*, volume 6, pages 459–470. World Scientific Press, 2001.

[44] K. Schmidt. LoLA: A low level analyser. In M. Nielsen and Dan S., editors, *Application and Theory of Petri Nets, 21st International Conference (ICATPN 2000)*, volume 1825 of *Lecture Notes in Computer Science*, pages 465–474. Springer, 2000.

[45] R. Seger and E. G. Krebs. The mapk signaling cascade. *FASEB Journal*, 9(9):726–735, 1995.

[46] M.-O. Stehr. A rewriting semantics for algebraic nets. In C. Girault and R. Valk, editors, *Petri Nets for System Engineering: A Guide to Modelling, Verification, and Applications*. Springer-Verlag, 2000.

[47] C. Talcott. Formal executable models of cell signaling primitives. In T. Margaria, A. Philippou, and B. Steffen, editors, *2nd International Symposium on Leveraging Applications of Formal Methods, Verification and Validation ISOLA06*, pages 303–307, 2006.

[48] C. Talcott. Symbolic modeling of signal transduction in pathway logic. In L. F. Perrone, F. P. Wieland, J. Liu, B. G. Lawson, D. M. Nicol, and R. M. Fujimoto, editors, *2006 Winter Simulation Conference*, pages 1656–1665, 2006.

[49] C. Talcott and D. L. Dill. Multiple representations of biological processes. *Transactions on Computational Systems Biology*, 2006.

[50] C. Talcott, S. Eker, M. Knapp, P. Lincoln, and K. Laderoute. Pathway logic modeling of protein functional domains in signal transduction. In *Proceedings of the Pacific Symposium on Biocomputing*, January 2004.

[51] A. Tiwari. Abstractions for hybrid systems. *Formal Methods in Systems Design*, 32(1):57–83, 2008.

[52] A. Tiwari, C. Talcott, M. Knapp, P. Lincoln, and K. Laderoute. Analyzing pathways using sat-based approaches. In H. Anai, K. Horimoto, and T. Kutsia, editors, *Algebraic Biology 2007*, volume 4545 of *Lecture Notes in Computer Science*, pages 155–169, 2007.

[53] I. Zevedei-Oancea and S. Schuster. Topological analysis of metabolic networks based on Petri net theory. *Silico Biology*, 3(0029), 2003.

Chapter

2

Probabilistic Model Checking for Systems Biology

Marta Kwiatkowska, Gethin Norman, and David Parker

Oxford University Computing Laboratory
Wolfson Building, Parks Road, Oxford, OX1 3QD

Abstract Probabilistic model checking is a technique for formally verifying quantitative properties of systems that exhibit stochastic behavior. In this chapter, we show how this approach can be applied to the study of biological systems such as biochemical reaction networks and signaling pathways. We present an introduction to the state-of-the-art probabilistic model-checking tool PRISM using a case study based on the Fibroblast Growth Factor (FGF) signaling pathway.

2.1 Introduction

Probabilistic model checking is a formal verification technique for the analysis of systems with stochastic characteristics. It is based on the exhaustive construction and analysis of a probabilistic model, typically a Markov chain or Markov process, of some real-life system. This approach allows various quantitative properties of the system's behavior, such as performance or reliability, to be analyzed automatically under a range of different scenarios or parameters. The results can then be used to identify anomalies, faults, or other points of interest in the systems being modeled. Probabilistic model checking has been used to study a wide array of systems, including randomized communication protocols such as Bluetooth and Firewire, security protocols for anonymity and contract signing, dynamic power management schemes, and NAND multiplexing for nanotechnology.

In this chapter, we focus on how probabilistic model checking can be used to study the behavior of biological systems such as biochemical reaction networks and signaling pathways that under certain assumptions, are well suited to being modeled as discrete stochastic systems. We give an introduction to probabilistic model checking and to the software tool PRISM [10, 17] that implements these techniques and, using a selection of examples of increasing complexity, demonstrate how PRISM can be used as a framework for the modeling and analysis of biological pathways. The approaches introduced in this chapter should not be viewed as a substitute for the classical approaches based on simulation and differential equations. Rather, these techniques should be used in conjunction with these classical approaches to obtain a more detailed understanding of the complex interactions and behavior of biological pathways.

The chapter is organized as follows. Section 2.2 presents an overview of probabilistic model checking, including descriptions of continuous-time Markov chains, the temporal logic CSL, and the probabilistic model-checking tool PRISM. Sections 2.3, 2.4, and 2.5 present three examples to demonstrate both how to model biological pathways in the PRISM language and how probabilistic model checking can be used to analyze such pathways. In Section 2.3, we start with a simple set of reactions. In Sections 2.4 and 2.5, we extend this simple example, first by increasing the number of components of each molecular species in the system, and then by extending the possible reactions of the system. Finally, Section 2.6 outlines related topics of interest.

2.2 Probabilistic Model Checking and PRISM

Probabilistic model checking is a technique for the modeling and analysis of systems that exhibit stochastic behavior. This technique is a variant of *model checking*, a well-established and widely used formal method for ascertaining the correctness of real-life systems. Model checking requires two inputs: (1) a description of the system in some high-level modeling formalism (such as a Petri net or process algebra), and (2) specification of one or more desired properties of that system, usually in temporal logic (e.g., CTL or LTL). From the former, a model of the system is constructed, typically a labeled state-transition system in which each state represents a possible system configuration and the transitions represent the evolution of the system from one configuration to another over time. It then becomes possible to automatically verify whether each property is satisfied, based on a systematic and exhaustive exploration of the model.

In probabilistic model checking, the models are augmented with quantitative information regarding the likelihood that transitions will occur and the times at which they will do so. In practice, these models are typically Markov chains or Markov decision processes. In this chapter, we use *continuous-time Markov chains* (CTMCs), in which transitions between states are assigned (positive, real-valued) rates, which are interpreted as the rates of negative exponential distributions. Properties, while still expressed in temporal logic, are now quantitative in nature. For example, rather than verifying that "the protein always eventually degrades," we may ask, "What is the probability that the protein eventually degrades?" or "What is the probability that the protein degrades within t hours?" The temporal logic most commonly used for this purpose is Continuous Stochastic Logic (CSL). Furthermore, by adding rewards to a CTMC, we can also specify properties such as "What is the expected energy dissipation through protein binding within the first t time units?" and "What is the expected number of binding reactions before relocation occurs?"

In the remainder of this section, we present an introduction to CTMCs, CSL, and the software tool PRISM, which provides support for probabilistic model checking of CTMCs using CSL. For further details, see sources such as [13].

2.2.1 Continuous-Time Markov Chains

Continuous-time Markov chains (CTMCs), which are frequently used in performance analysis, model both continuous real-time and probabilistic choice. This is done by specifying the *rate* at which a transition between two states occurs. Formally, we have the following definition:

Definition 2.1

A CTMC is a tuple $\mathcal{C} = (S, \mathbf{R}, L)$ where S is a finite set of states; $\mathbf{R} : S \times S \to \mathbb{R}_{\geq 0}$ is a transition rate matrix; $L : S \to 2^{AP}$ is a labeling function.

The transition rate matrix $\mathbf{R}$ assigns rates to each pair of states, which are used as the parameter of an exponential distribution. A transition can only occur between states s and s' if $\mathbf{R}(s, s'){>}0$ and, in this case, the probability of this transition being triggered within t time units equals $1 - e^{-\mathbf{R}(s,s')\cdot t}$. Typically, in a state s, there is more than one state s' for which $\mathbf{R}(s, s'){>}0$; this is known as a *race condition* and the first transition to be triggered determines the next state. The choice of successor state s' from state s is thus probabilistic. The probability of moving to s' is $\frac{\mathbf{R}(s,s')}{E(s)}$, where $E(s) = \sum_{s' \in S} \mathbf{R}(s, s')$ is the *exit rate* of state s. The total time spent in s before any transition occurs is exponentially distributed with rate $E(s)$. The labeling function L assigns *atomic propositions* from a set AP to each state of the CTMC. These are used to label states with properties of interest.

A CTMC can be augmented with *reward structures* that are used to annotate its states and/or transitions with additional quantitative information. Formally, a reward structure for a CTMC is a pair $(\underline{\rho}, \iota)$ where

- $\underline{\rho} : S \rightarrow \mathbb{R}_{\geq 0}$ is the *state reward function*;
- $\iota : S \times S \rightarrow \mathbb{R}_{\geq 0}$ is the *transition reward function*.

A path of a CTMC $\mathcal{C} = (S, \mathbf{R}, L)$ is a nonempty sequence $s_0 t_0 s_1 t_1 s_2 \ldots$ where $s_i \in S$, $t_i \in \mathbb{R}_{>0}$ and $\mathbf{R}(s_i, s_{i+1}) > 0$ for all $i \geq 0$. The value t_i represents the amount of time spent in the state s_i. We denote by $\omega@t$ the state occupied at time t; that is, s_j, where j is the smallest index for which $\sum_{i=0}^{j} t_i \geq t$. We denote by $Path(s)$ the set of all (infinite and finite) paths of the CTMC $\mathcal{C}$ starting in state s. A probability measure Pr_s over $Path(s)$ can then be derived [2].

Two traditional properties of CTMCs are *transient* behavior, which relates to the state of the model at a particular time instant, and *steady-state* behavior, which describes the state of the CTMC in the long run. For a CTMC $\mathcal{C} = (S, \mathbf{R}, L)$, the transient probability $\pi_{s,t}(s')$ is defined as the probability, having started in state s, of being in state s' at time instant t. The steady-state probability $\pi_s(s')$ is the probability of, having started in state s, being in state s' in the long run. The steady-state probability distribution [i.e., the values $\pi_s(s')$ for all $s' \in S$,] can be used to infer the percentage of time, in the long run, that the CTMC spends in each state.

2.2.2 Continuous Stochastic Logic

The temporal logic CSL originally introduced by Aziz et al. [1] and since extended by Baier et al. [2] is based on the temporal logics CTL [5] and PCTL [7] and provides a powerful means to specify both path-based and traditional state-based performance measures on CTMCs. We use an extended version [13] that also allows for the specification of reward properties.

Definition 2.2 The syntax of CSL is as follows:

$$\phi \ ::= \ \mathtt{true} \mid a \mid \neg\phi \mid \phi \wedge \phi \mid \mathtt{P}_{\sim p}[\phi \ \mathtt{U}^{I} \ \phi] \mid \mathtt{S}_{\sim p}[\phi] \mid \mathtt{R}_{\sim r}[\mathtt{I}^{=t}] \mid \mathtt{R}_{\sim r}[\mathtt{C}^{\leq t}] \mid \mathtt{R}_{\sim r}[\mathtt{F} \ \phi] \mid \mathtt{R}_{\sim r}[\mathtt{S}]$$

where a is an atomic proposition, $\sim \in \{<, \leq, \geq, >\}$, $p \in [0, 1]$, I is an interval of $\mathbb{R}_{\geq 0}$, and $r, t \in \mathbb{R}_{\geq 0}$.

CSL formulae are evaluated over the states of a CTMC, and we write $s \models \phi$ to indicate that the CSL formula ϕ is true to state s. We also say that ϕ *holds* or *is satisfied* in s. CSL includes the standard operators from propositional logic: `true` (satisfied in all states); atomic propositions (a is true in states that are labeled with a); negation ($\neg\phi$ is true if ϕ is not); and conjunction ($\phi_1 \wedge \phi_2$ is true if both ϕ_1 and ϕ_2 are true). From these, we can derive other standard Boolean operators such as disjunction ($\phi_1 \vee \phi_2 \equiv \neg(\neg\phi_1 \wedge \neg\phi_2)$) and implication ($\phi_1 \Rightarrow \phi_2 \equiv \neg\phi_1 \vee \phi_2$) in the usual way.

CSL also includes two probabilistic operators, P and S, both of which include a probability bound $\sim p$. A formula $\mathtt{P}_{\sim p}[\psi]$ is true in a state s if the probability of the path formula ψ being satisfied from state s meets the bound $\sim p$. In this chapter we use a single type of path formula, $\phi_1\ \mathtt{U}^I\ \phi_2$, called an *until* formula, which is true of a path ω if, for some time instant $t \in I$, at time t in the path ω the CSL subformula ϕ_2 is true and the subformula ϕ_1 is true at all preceding time instants. Perhaps the most common use of this operator is the case where ϕ_1 is `true`, in which case $\mathtt{P}_{\sim p}[\mathtt{true}\ \mathtt{U}^I\ \phi]$ states that the probability of ϕ being true at some point in the interval I satisfies $\sim p$. In particular, this allows reasoning about transient behavior of a CTMC ($\mathtt{P}_{\sim p}[\mathtt{true}\ \mathtt{U}^{[t,t]}\ \phi]$ that is, the probability of ϕ being true at time instant t)—and untimed reachability ($\mathtt{P}_{\sim p}[\mathtt{true}\ \mathtt{U}^{[0,\infty)}\ \phi]$, the probability of ϕ eventually being true). In the latter case we often omit the interval, so that $\mathtt{P}_{\sim p}[\mathtt{true}\ \mathtt{U}\ \phi] \equiv \mathtt{P}_{\sim p}[\mathtt{true}\ \mathtt{U}^{[0,\infty)}\ \phi]$. The S operator is used to specify the steady-state behavior of a CTMC. More precisely, $\mathtt{S}_{\sim p}[\phi]$ asserts that the steady-state probability of being in a state satisfying ϕ meets the bound $\sim p$.

CSL also has an R operator for properties concerning the expected value of rewards. The formula $\mathtt{R}_{\sim r}[\mathtt{I}^{=t}]$ asserts that the expected value of the state reward at time instant t meets the bound $\sim r$. Similarly, $\mathtt{R}_{\sim r}[\mathtt{C}^{\leq t}]$ refers to the expected reward accumulated up until time t, $\mathtt{R}_{\sim r}[\mathtt{F}\ \phi]$ to the expected reward accumulated before a state satisfying ϕ is reached, and $\mathtt{R}_{\sim r}[\mathtt{S}]$ to the long-run average expected reward. Note that the $\mathtt{I}^{=t}$ operator refers only to state rewards, whereas the others refer to both state and transition rewards. Also, in the latter case, state rewards are interpreted as the rate at which rewards are accumulated; that is, spending time t in state s accumulates a reward of $\rho(s)\cdot t$.

The semantics of CSL over CTMCs is defined as follows:

Definition 2.3 Let $\mathcal{C} = (S, \mathbf{R}, L)$ be a labeled CTMC. For any state $s \in S$, the relation $s \models \phi$ is defined inductively by

$$
\begin{array}{lll}
s \models \mathtt{true} & \text{for all } s \in S & \\
s \models a & \Leftrightarrow & a \in L(s) \\
s \models \neg\phi & \Leftrightarrow & s \not\models \phi
\end{array}
$$

$$
\begin{array}{rcl}
s \models \phi_1 \wedge \phi_2 & \Leftrightarrow & s \models \phi_1 \wedge s \models \phi_2 \\
s \models \mathtt{P}_{\sim p}[\psi] & \Leftrightarrow & Prob(s, \psi) \sim p \\
s \models \mathtt{S}_{\sim p}[\phi] & \Leftrightarrow & \sum_{s' \models \phi} \pi_s(s') \sim p \\
s \models \mathtt{R}_{\sim r}[\mathtt{I}^{=t}] & \Leftrightarrow & Exp(s, X_{\mathtt{I}^{=t}}) \sim r \\
s \models \mathtt{R}_{\sim r}[\mathtt{C}^{\leq t}] & \Leftrightarrow & Exp(s, X_{\mathtt{C}^{\leq t}}) \sim r \\
s \models \mathtt{R}_{\sim r}[\mathtt{F}\ \phi] & \Leftrightarrow & Exp(s, X_{\mathtt{F}\phi}) \sim r \\
s \models \mathtt{R}_{\sim r}[\mathtt{S}] & \Leftrightarrow & \lim_{t \to \infty} \frac{1}{t} \cdot Exp(s, X_{\mathtt{C}^{\leq t}}) \sim r
\end{array}
$$

where $Prob(s, \psi) \stackrel{\text{def}}{=} Pr_s\{\omega \in Path(s) \mid \omega \models \psi\}$, $Exp(s, X)$ denotes the expectation of the random variable X with respect to the probability measure Pr_s, and for any path $\omega = s_0 t_0 s_1 t_1 s_2 \cdots \in Path(s)$

$$
\omega \models \phi_1\ \mathtt{U}^I\ \phi_2 \Leftrightarrow \exists t \in I \cdot (\omega@t \models \phi_2 \wedge \forall x \in [0, t) \cdot (\omega@x \models \phi_1))
$$

$$
X_{\mathtt{I}^{=t}}(\omega) = \underline{\rho}(\omega@t)
$$

$$
X_{\mathtt{C}^{\leq t}}(\omega) = \sum_{i=0}^{j_t - 1} (t_i \cdot \underline{\rho}(s_i) + \iota(s_i, s_{i+1})) + \left(t - \sum_{i=0}^{j_t - 1} t_i\right) \cdot \underline{\rho}(s_{j_t})
$$

$$
X_{\mathtt{F}\phi}(\omega) = \begin{cases} 0 & \textit{if } \omega(0) \models \phi \\ \infty & \textit{if } \forall i \in \mathbb{N}.\, s_i \not\models \phi \\ \sum_{i=0}^{\min\{j \mid s_j \models \phi\} - 1} t_i \cdot \underline{\rho}(s_i) + \iota(s_i, s_{i+1}) & \text{otherwise} \end{cases}
$$

and $j_t = \min\{j \mid \sum_{i=0}^{j} t_i \geq t\}$.

In many cases, it is more useful to generate quantitative, rather than Boolean, results for CSL properties. For this purpose, we allow the bounds $\sim p$ and $\sim r$ attached to the P, S, and R operators to be replaced with =?. This permits, for example, properties such as the following:

- $\mathtt{P}_{=?}[\mathtt{true}\ \mathtt{U}^{[t,t]}\ \phi]$; "What is the probability of ϕ being true at time t?"
- $\mathtt{S}_{=?}[\phi]$; "What is the long-run probability of ϕ holding?"
- $\mathtt{R}_{=?}[\mathtt{F}\ \phi]$; "What is the expected reward accumulated before a state satisfying ϕ is reached?"

The =? form of the P, S, and R operators can be used only when it is the outermost operator in a CSL formula.

2.2.3 PRISM

PRISM [10, 17] is a probabilistic model checker that provides automatic verification of CTMCs using the logic CSL, as described in the previous sections. It also provides support for two other types of probabilistic models: discrete-time Markov chains and Markov decision processes. See references such as [22] for more information.

The PRISM tool accepts probabilistic models described in the *PRISM modeling language*, a simple, state-based language that will be explained in the later sections of this chapter. The basic functionality of PRISM is to take a model described in this formalism, construct the corresponding probabilistic model (in this case, a CTMC), and then perform model checking of one or more CSL properties. PRISM also supports the notion of *experiments*, which are a way of automating multiple instances of model checking. Experiments allow the user to easily obtain the outcome of one or more properties as functions of model and property parameters. The resulting table of values can either be viewed directly, exported for use in another application such as a spreadsheet, or displayed using PRISM's graph-plotting tool. This is frequently a good way of identifying interesting patterns or trends in the behavior of a system. You will see several examples of this throughout this chapter.

Figure 2.1 shows a screenshot of the PRISM graphical user interface, illustrating the results of a model-checking experiment being plotted on a graph. The tool also features a built-in text editor for the PRISM language and a discrete-event simulation engine that can be used to debug PRISM models. Alternatively, all model-checking functionality is available in a command-line version of the tool. PRISM is a *free*, *open-source* application. It presently operates on Linux, UNIX, Windows, and Macintosh operating systems. Both binary and source code versions can be downloaded from the website [17].

PRISM incorporates a variety of techniques to construct CTMCs and perform CSL model checking on them. Much of the underlying work is a combination of graph-theoretical algorithms (e.g., for reachability analysis)

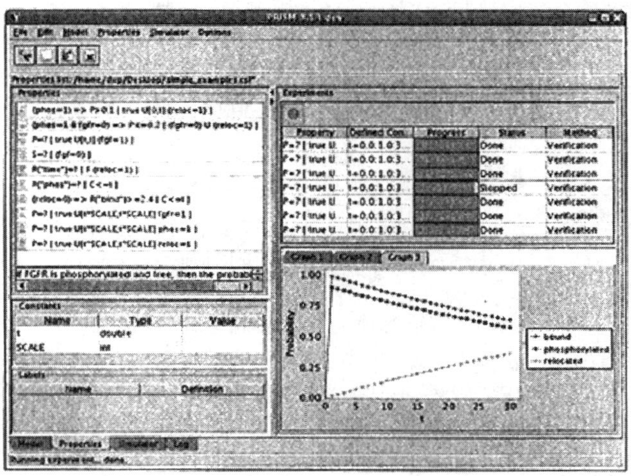

Figure 2.1 A screenshot of the PRISM graphical user interface

and numerical computation (e.g., to calculate probabilities and expected reward values for each state of the CTMC). For the latter, due to the sizes of the problems typically solved, PRISM uses iterative numerical solution methods. Some aspects of CSL model checking (e.g., those based on long-run properties of a CTMC) require the solution of linear equation systems. For these operations, well-known techniques such as the Jacobi, Gauss–Seidel, and SOR (successive over-relaxation) methods are available. For timed operators of CSL (i.e., those based on transient CTMC properties), PRISM implements another common iterative numerical method called uniformization. In addition, the tool provides approximate solution methods based on Monte Carlo techniques and the built-in discrete-event simulation engine.

One final notable aspect of PRISM is that it is a *symbolic* model checker: It is implemented primarily using data structures based on binary decision diagrams (BDDs)—for example, *multi-terminal* binary decision diagrams (MTBDDs) and other variants. These data structures provide compact representations and efficient manipulation of large, structured probabilistic models by exploiting regularity from the high-level descriptions of the models. To maximize its efficiency, PRISM actually uses combinations of these symbolic data structures and conventional *explicit* storage schemes such as sparse matrices and arrays. See sources such as [12, 15] for more information about the implementation of PRISM.

2.3 Modeling a Simple Biological System in PRISM

We now illustrate how PRISM can be used as a tool for modeling and analyzing biological systems. We will do this by developing a case study based on the role of fibroblast growth factor (FGF) in receptor biosynthesis. The FGF family of proteins plays an important role in cell signaling—for example, in wound healing.

2.3.1 A Simple Set of Reactions

Figure 2.2 shows a simple set of reactions based on the role of FGF in receptor biosynthesis, which can be summarized as follows: An FGF ligand (molecule) can bind to an FGF receptor (FGFR) to form the compound FGF:FGFR. Once the compound FGF:FGFR is formed, FGFR can become phosphorylated. These two reactions (binding and phosphorylation) are also reversible. Finally, when FGFR is phosphorylated, it can be relocated (we assume that FGF disappears if it is bound to FGFR when relocation occurs). Each reaction has an associated *kinetic rate* (values k_1 to k_5 in Figure 2.2). Note that for the binary reaction (binding), the units of the kinetic rate include the concentrations of the reactants. More precisely,

1: FGF binds/releases FGFR	
FGF + FGFR $\rightarrow$ FGF:FGFR	$k_1 = 5e + 8\ M^{-1}s^{-1}$
FGF + FGFR $\leftarrow$ FGF:FGFR	$k_2 = 0.002\ s^{-1}$
2: Phosphorylation of FGFR (in the form of FGF:FGFR)	
FGFR $\rightarrow$ FGFRP	$k_3 = 0.1\ s^{-1}$
3: Dephosphorylation of FGFR	
FGFRP $\rightarrow$ FGFR	$k_4 = 0.01\ s^{-1}$
4: Relocation of FGFR (in the form of FGFRP)	
FGFR $\rightarrow$ relocFGFR	$k_5 = 1/60\ min^{-1}$

Figure 2.2 Summary of the reactions

the kinetic rate k_1 has units $M^{-1}s^{-1}$, where M refers to the *molar concentration* (i.e., the number of moles per liter).

Our approach to modeling the behavior of biochemical reaction systems is as follows: We consider a collection of molecules, from a variety of molecular species, interacting according to a set of reactions. We make the assumption that the reactions take place in a spatially uniform mixture in a fixed volume at constant pressure and temperature. Under this assumption, two distinct modeling approaches have been proposed. The first, which we refer to as the *continuous deterministic* approach, represents the number of molecules of each species at time t by a continuous function. A set of ordinary differential equations is then derived, the solutions to which give the (average) concentration of each molecular species over time.

The second approach, and the one used in this chapter, is to use a *discrete stochastic* model. In this case, the amount of each molecular species is modeled as a discrete quantity and the occurrence of a reaction between one or more of these molecules is considered a discrete event. The evolution of this model over time is inherently stochastic. In fact, the underlying model can be shown to be a continuous-time Markov chain, where the stochastic rates associated with each transition of the CTMC can be derived from the kinetic rates of the reaction system. In the case of unary reactions, the stochastic rate equals the kinetic rate. For binary reactions, supposing the kinetic rate is given in terms of molar concentrations, the stochastic rate is obtained by dividing by $V \cdot \mathcal{N}$, where V is the volume and $\mathcal{N}$ is Avogadro's number. For a more in-depth coverage of this topic, see, for example, [6, 24].

We now return to the simple example described previously. The different species that can be present in the system are as follows: free FGF (FGF), free FGFR (FGFR), FGF bound to FGFR (FGF:FGFR), FGF bound to phosphorylated FGFR (FGF:FGFRP), free phosphorylated FGFR (FGFRP), and relocated FGFR (relocFGFR). Making the assumption that the system initially comprises a single FGF ligand and a single FGFR receptor, the corresponding CTMC for this example is given in Figure 2.3, where the stochastic rates ($\mathbf{k_i}$) are derived from the kinetic rates (k_i) via the calculation described in the previous paragraph.

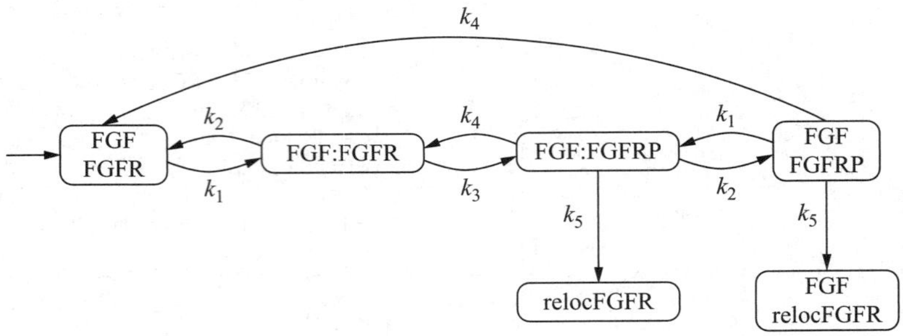

Figure 2.3 Underlying CTMC of the system of reactions given in Figure 2.2

2.3.2 PRISM Models

We now give two alternative approaches for modeling these reactions in PRISM, shown in Figures 2.4 and 2.5, respectively. A model described in the PRISM language comprises a keyword, corresponding to the model type (`ctmc` in this case), and a set of *modules*, the state of each being represented by a set of finite-ranging *variables*. The behavior of this module (i.e., the changes in states that it can undergo) is specified by a number of *guarded commands* of the form $[]\ g \rightarrow r : u$. The interpretation of a command is that if the predicate (guard) g is true, then the system is updated according to u, which comprises one or more statements of the form $(x' = \ldots)$ indicating how the value of variable x is changed. The rate at which this occurs is given by r; that is, this is the value that will be attached to the corresponding transition in the underlying CTMC.

PRISM supports *synchronization* between modules in the style of process algebras. This is achieved by labeling commands with *actions* (placed between the initial square brackets). Transitions in different modules labeled with the same action occur simultaneously. The rate of synchronized transitions is equal to the product of the individual rates of the commands of the different modules that are synchronized. Because the product of several rates is not always meaningful, a commonly used technique, as seen here, is to make one action *active*, which actually defines the rate for the synchronized transition, and the others *passive*, with rate 1. In PRISM, when a rate is omitted for a command, it is assumed to be 1. By default, all modules are combined using the standard CSP parallel composition [21] (i.e., modules synchronize over all their common actions). In addition, by using the `system ... endsystem` construct, several other CSP operators, including asynchronous parallel composition and hiding, can be employed when combining PRISM modules. Some of these constructs will be illustrated in examples later in this chapter. For further details, see the PRISM manual [17].

```
ctmc

const double k1 = 5000; // rate of binding
const double k2 = 0.002; // rate of release
const double k3 = 0.1; // rate of phosphorylation
const double k4 = 0.01; // rate of dephosphorylation
const double k5 = 1/(60 * 60); // rate of relocation

module FGF

    fgf : [0..2] init 0; // 0 - free, 1 - bound, 2 - removed from system

    [bind]  fgf=0 → (fgf'=1); // FGF and FGFR bind
    [rel]   fgf=1 → (fgf'=0); // FGF and FGFR unbind
    [reloc] fgf=1 → (fgf'=2); // FGF disappears since bound when FGFR relocates

endmodule

module FGFR

    fgfr  : [0..1] init 0; // 0 - free, 1 - bound
    phos  : [0..1] init 0; // 0 - unphosphorylated, 1 - phosphorylated
    reloc : [0..1] init 0; // 0 - not relocated, 1 - relocated

    [bind] fgfr=0 & reloc=0 → (fgfr'=1); // FGF and FGFR bind
    [rel] fgfr=1 & reloc=0 → (fgfr'=0); // FGF and FGFR unbind
    [] fgfr=1 & phos=0 & reloc=0 → k3 : (phos'=1); // FGFR phosphorylates
    [] phos=1 & reloc=0 → k4 : (phos'=0); // FGFR dephosphorylates
    [] phos=1 & fgfr=0 & reloc=0 → k5 : (reloc'=1) & (fgfr'=0) & (phos'=0); // relocates
    [reloc] phos=1 & fgfr=1 & reloc=0 → (reloc'=1) & (fgfr'=0) & (phos'=0); // relocates

endmodule

module RATES

    [bind] true → k1 : true; // FGF and FGFR bind
    [rel] true → k2 : true; // FGF and FGFR unbind
    [reloc] true → k5 : true; // FGFR relocates

endmodule
```

Figure 2.4 First possible PRISM representation

In our first approach (Figure 2.4), we represent the two main molecules (FGF and FGFR) as separate modules, each with variables representing its current state (details are given in the comments after each variable declaration). Reactions involving more than one species are modeled using synchronization. Consider, for example, the binding of FGF and FGFR, modeled by the first commands of modules *FGF* and *FGFR* and synchronization on the action label *bind*. In this case, the variables *fgf* and *fgfr* in the two modules simultaneously change from 0 to 1. For this model, we use a third module *RATES* to store the rates of synchronous transitions. This module also appears in the synchronization on *bind*; thus the overall rate is *k1*. Reactions involving only a single protein can be modeled without synchronization, as seen in the third, fourth, and fifth commands of module *FGFR*.

```
ctmc

const double k1 = 5000; // rate of binding
const double k2 = 0.002; // rate of release
const double k3 = 0.1; // rate of phosphorylation
const double k4 = 0.01; // rate of dephosphorylation
const double k5 = 1/(60 * 60); // rate of relocation

module SYSTEM

    x : [0..5] init 0;
    // 0 - FGF and FGFR free
    // 1 - FGF and FGFR free (FGFR phosphorylated)
    // 2 - FGF and FGFR bound
    // 3 - FGF and FGFR bound (FGFR phosphorylated)
    // 4 - FGF free and FGFR relocated
    // 5 - FGFR relocated

    // FGF and FGFR bind
    [] x=0 → k1 : (x'=2); // FGFR not phosphorylated
    [] x=1 → k1 : (x'=3); // FGFR phosphorylated
    // FGF and FGFR unbind
    [] x=2 → k2 : (x'=0); // FGFR not phosphorylated
    [] x=3 → k2 : (x'=1); // FGFR phosphorylated
    // FGFR becomes phosphorylated (FGF must be bound)
    [] x=2 → k3 : (x'=3);
    // FGFR dephosphorylates
    [] x=1 → k4 : (x'=0); // FGF not bound
    [] x=3 → k4 : (x'=2); // FGF bound
    // FGFR relocates (FGFR must be phosphorylated)
    [] x=1 → k5 : (x'=4); // FGF not bound
    [] x=3 → k5 : (x'=5); // FGF bound

endmodule
```

Figure 2.5 Second possible PRISM representation

In our second approach (Figure 2.5), we use a single module with one variable x representing the (six) possible states of the whole system. The meaning of each value of x is given in the comments, which correspond directly to the states of the CTMC shown in Figure 2.3. These two PRISM models result in identical CTMCs. The second is a more concise description in PRISM, but the first has a more intuitive state encoding and, therefore, is easier to modify and to express properties for. In general, a combination of these two modeling approaches is used: In simple cases it is possible to use a single variable, but as the system becomes more complex the use of separate variables and synchronization becomes more desirable. We will see this approach as we develop this case study further in Sections 2.4 and 2.5.

Finally, we add reward structures to our model. In PRISM, these are described using the following construct:

```
rewards "reward_name" ... endrewards
```

```
// time spent in any state
rewards "time"

    true : 1;

endrewards
```

```
// time spent phosphorylated
rewards "phos"

    phos=1 : 1;

endrewards
```

```
// number of bindings
rewards "bind"

    [bind] true : 1;

endrewards
```

Figure 2.6 Three reward structures for the PRISM model of Figure 2.4

They are specified using multiple reward items of the form

$$g\ :\ r;\quad \text{or}\quad [a]\ g\ :\ r;$$

to describe state and transition rewards, respectively. In the preceding syntax, g is a predicate (over all the variables of the model), a is an action label appearing in the commands of the model, and r is a real-valued expression (which can contain any variables, constants, or other items from the model). A reward item "$g\ :\ r$" assigns a state reward of r to all states satisfying g; a reward item "$[a]\quad g\ :\ r$" assigns a transition reward of r to all a-labeled transitions from states satisfying g. Multiple rewards (from different reward items) for a single state or transition are summed, and states or transitions with no assigned reward are assumed to have reward 0.

Figure 2.6 shows three such reward structures for the PRISM model of Figure 2.4. The first reward structure ("*time*") simply assigns a state reward of 1 to all states in the model. It can be used, for example, to analyze the total expected time before some event occurs. The second reward structure ("*phos*") assigns a state reward of 1 only to those states in which FGFR is phosphorylated. It could be used to compute the amount of time that FGFR spends phosphorylated within a particular period of time or the expected amount of phosphorylated FGFR at a specific time instant. Lastly, the reward structure "*bind*" assigns a reward of 1 to all transitions that correspond to a binding between FGF and FGFR.

2.3.3 Model Analysis with PRISM

We now demonstrate how the temporal logic CSL (see Section 2.2.2) can be used to express properties of the reaction system model presented in Figure 2.4, augmented with the reward structures from the previous section. Following are some CSL examples, along with their informal meaning. Because multiple reward structures exist, we will adopt PRISM's notation R{"*rew*"} to indicate use of the CSL reward operator R using the reward structure with the name "*rew*." Note also that, in our model, the unit of time used is seconds.

- $(phos{=}1) \Rightarrow \mathtt{P}_{>0.1}[\ \mathtt{true}\ \mathtt{U}^{[0,t]}\ (reloc{=}1)\ ]$; "If FGFR is currently phosphorylated, then the probability of it being relocated within the next t seconds is greater than 0.1."

- $(phos{=}1 \wedge fgfr{=}0) \Rightarrow \mathtt{P}_{\leq 0.2}[\ (fgfr{=}0)\ \mathtt{U}\ (reloc{=}1)\ ]$; "If FGFR is phosphorylated and free, then the probability of it being relocated before binding to FGF is at most 0.2."
- $\mathtt{P}_{=?}[\ \mathtt{true}\ \mathtt{U}^{[t,t]}\ (fgf{=}1)\ ]$; "The probability that FGF is bound to FGFR at time instant t (i.e., after exactly t seconds)."
- $\mathtt{P}_{=?}[\ \mathtt{true}\ \mathtt{U}\ (reloc{=}1 \wedge fgf{=}2)\ ]$; "The probability that FGFR relocates and FGF is bound when relocation occurs."
- $\mathtt{S}_{=?}[\ (fgf{=}0)\ ]$; "The probability that, in the long run, FGF is free."
- $\mathtt{R}\{\text{"}time\text{"}\}_{=?}[\ \mathtt{F}\ (reloc{=}1)\ ]$; "The expected time taken before FGFR relocates."
- $\mathtt{R}\{\text{"}phos\text{"}\}_{=?}[\ \mathtt{C}^{\leq t}\ ]$; "The expected time that FGFR spends phosphorylated within the next t seconds."
- $(reloc{=}0) \Rightarrow \mathtt{R}\{\text{"}bind\text{"}\}_{\geq 2.4}[\ \mathtt{C}^{\leq t}\ ]$; "If FGFR is not relocated, the expected number of bindings during the next t seconds is at least 2.4."

When analyzing quantitative properties of a system such as these, it is often useful to study how the properties vary as changes are made to parameters either in the properties (e.g., t) or in the model. Analysis of this kind is much more likely to provide insight into the model or to identify interesting or anomalous behavior.

To illustrate this point, Figure 2.7 shows results obtained with PRISM for the probabilities that, at time instant t, FGFR is (1) bound to FGF, (2) phosphorlylated, or (3) relocated. The first of these, for example, uses the CSL property $\mathtt{P}_{=?}[\ \mathtt{true}\ \mathtt{U}^{[t,t]}\ (fgf{=}1)\ ]$ from the list given earlier. Results are plotted for ranges of t over three different time scales (seconds, minutes, and hours). Figure 2.7(a) shows that in the initial evolution of the system, FGF and FGFR bind very quickly (and remain bound), after which FGFR becomes phosphorylated while there is almost no chance of FGFR relocating. Figures 2.7(b) and 2.7(c) show, however, that as time elapses, the chance that FGF and FGFR are bound diminishes and the chance that FGFR is phosphorylated diminishes even more rapidly. In addition, we see that eventually FGFR will become relocated.

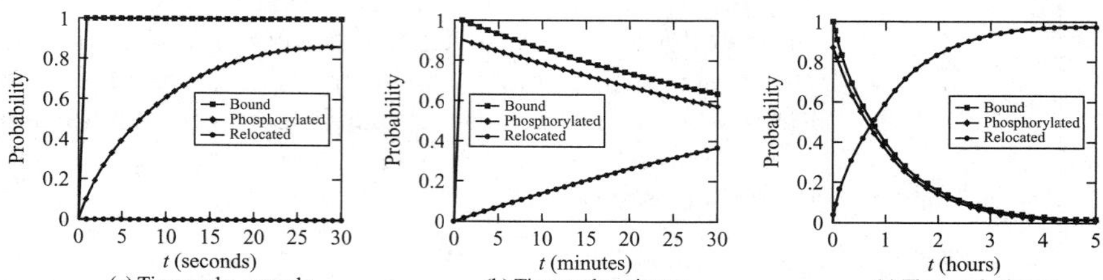

(a) Time scale: seconds (b) Time scale: minutes (c) Time scale: hours

Figure 2.7 Transient properties of FGFR for the model of Figure 2.4

EXERCISES

1. Based on the model in Figure 2.4 and the reward structures in Figure 2.6, write CSL specifications for the following properties:

(a) If FGFR is currently phosphorylated, then the probability that it remains phosphorylated until relocation occurs is at most 0.65
(b) The probability that FGFR is phosphorylated at time instant t
(c) The probability that FGFR is phosphorylated for the first time within the time interval $[t_1, t_2]$
(d) The expected time that FGFR is phosphorylated before it relocates
(e) The expected number of times that FGF and FGFR bind before FGFR relocates

2. Construct appropriate reward structures for properties of the model in Figure 2.4 relating to the expected time that FGF and FGFR are bound and the expected number of bindings *and* unbindings. Write CSL specifications for calculating the expected time that FGF and FGFR spend bound during the first t seconds, the expected number of bindings and unbindings in this time, and the expected time spent bound before relocation occurs.

3. Extend the model of Figure 2.4 with a variable to count the number of phosphorylations and write CSL specifications for the properties:

(a) The probability that at least l phosphorylations occur within the first t seconds
(b) The probability that at most l phosphorylations occur before relocation

Hint: Because the variable you add to the model must be bounded, make sure that the bound is larger than required for the property.

4. Based on the model of Figure 2.4, write a new version in which each of the six possible species (FGF, FGFR, FGF:FGFR, FGF:FGFRP, FGFRP, relocFGFR) is represented by a separate PRISM module. Check that the states and transitions in the new model match those of the original one and that numerical results such as those in Figure 2.7 agree.
Hint: In this model, the relevant modules will need to synchronize when phosphorylation and dephosphorylation occur.

2.4 Modeling Larger Numbers of Molecules

2.4.1 Individual-Based Representations

FGF and FGFR molecules (and thus molecules of the other various species) are known to occur in the system. One possible approach to representing this situation is to model each individual molecule in the system separately.

```
ctmc

const double N = 3; // number of each type

const double k1 = 5000/N; // rate of binding
const double k2 = 0.002; // rate of release
const double k3 = 0.1; // rate of phosphorylation
const double k4 = 0.01; // rate of dephosphorylation
const double k5 = 1/(60 * 60); // rate of relocation

module FGF .... endmodule

// construct further FGF molecules through renaming
module FGF2 = FGF [fgf=fgf2] endmodule
module FGF3 = FGF [fgf=fgf3] endmodule

module FGFR .... endmodule

// construct further FGF molecules through renaming
module FGFR2 = FGFR [fgfr=fgfr2, phos=phos2, reloc=reloc2] endmodule
module FGFR3 = FGFR [fgfr=fgfr3, phos=phos3, reloc=reloc3] endmodule

module RATES

    [bind] true → k1 : true; // FGF and FGFR bind
    [rel] fgf +fgf2 +fgf3>0 → k2/(fgf +fgf2 +fgf3) : true; // FGF and FGFR unbind
    [reloc] fgf +fgf2 +fgf3>0 → k2/(fgf +fgf2 +fgf3) : true; // FGFR relocates

endmodule

system // system definition (molecules of the same type do not interact)

    (FGF ||| FGF2 ||| FGF3) || (FGFR ||| FGFR2 ||| FGFR3) || RATES

endsystem
```

Figure 2.8 Individual-based PRISM representation (extending Figure 2.4)

This is sometimes necessary when the properties of interest for the system refer to the behavior of a particular molecule.

Figure 2.8 shows PRISM code that can be used to extend the previous model of Figure 2.4 to contain three of each molecule. First, we add a new PRISM module for each extra molecule. This task is done using a PRISM language feature called *renaming*; it creates a copy of an existing module, identical to the original except for its variable names, which are given new names (in fact, other constants and actions can also be renamed in the same way). This approach is equivalent to writing out each new module separately but is less error prone and improves the readability and scalability of the model.

Figure 2.8 also illustrates the use of PRISM's `system ... endsystem` construct, which describes how the modules in the system are composed to produce the full model. Because the FGF modules do not interact with each other, they are composed using the asynchronous parallel operator (|||), which stops them from synchronizing on any actions. The same is true for the FGFR modules. The (standard) parallel operator (||) is then used to compose these subsystems; in other words, we make these subsystems synchronize over their common actions. If the asynchronous parallel operator had not been used here, we would have needed to introduce a different action label for each possible interaction between FGF and FGFR (e.g., actions of the form *bind_i_j* for binding of the ith FGF molecule with the jth FGFR molecule) to stop FGF (or FGFR) molecules from reacting with one another. PRISM also supports several other CSP-based process-algebraic operators to aid modeling of complex systems; see the manual [17] for details.

We also need to modify the reaction rates of the system when increasing the number of molecules. More precisely, supposing that the volume of the system remains proportional to the initial number of FGF molecules, then the rates of binary reactions in Figure 2.8 (i.e., k_1) are obtained from those in Figure 2.4 by dividing by the initial number of FGF molecules (see the earlier discussion of computing stochastic reaction rates in Section 2.3.1).

2.4.2 Population-Based Representations

As might be expected, a potential problem with the model outlined previously is scalability. The individual-based modeling approach will suffer from the well-known state-space explosion problem. That is, as the complexity of the system under study increases, an exponential growth occurs in the state space of the underlying model.

An alternative is to employ a "population"-based approach where the number of each type of molecule or species is modeled, rather than the state of each individual component. In terms of the PRISM modeling language, this technique involves the use of variables as counters; that is, there is a counter for each of the possible species that can be present in the system (in this case, free FGF, free FGFR, free phosphorylated FGFR, FGF bound to FGFR, FGF bound to phosphorylated FGFR, and relocated free FGFR). In such a model, for example, an FGF ligand can bind with a FGFR receptor if the counters for the number of free FGF ligands and FGFR receptors are both greater than 0, and the occurrence of such a reaction is modeled by decrementing these counters and incrementing the counter representing the number of FGF ligands bound to FGFR receptors.

Figure 2.9 depicts a population-based PRISM model for the running example. It is based on the earlier model representing a single instance of each species (Figure 2.5). Note that we must adjust the rates of the CTMC to take into account the different possible interactions encoded. For example, if there are three FGF ligands (FGF_1, FGF_2, and FGF_3) and two FGFR receptors ($FGFR_1$ and $FGFR_2$), then six species are possible: FGF_1:$FGFR_1$, FGF_1:$FGFR_2$, FGF_2:$FGFR_1$, FGF_2:$FGFR_2$, FGF_3:$FGFR_1$, and FGF_3:$FGFR_2$. This is achieved by multiplying the rate

```
ctmc

const int N; // number of possible elements of FGF
const int M; // number of possible elements of FGFR
const int K = min (N, M); // number of possible elements of FGF:FGFR

const double k1 = 5000/N; // rate of binding
const double k2 = 0.002; // rate of release
const double k3 = 0.1; // rate of phosphorylation
const double k4 = 0.01; // rate of dephosphorylation
const double k5 = 1/(60 * 60); // rate of relocation

module POPULATION MODEL

    fgf : [0..N] init N; // free FGF
    fgfr : [0..M] init M; // free FGFR (not phosphorylated)
    fgfrp : [0..M] init 0; // free FGFR (phosphorylated)
    bnd : [0..K] init 0; // bound FGF:FGFR (FGFR not phosphorylated)
    bndp : [0..K] init 0; // bound FGF:FGFR (FGFR phosphorylated)
    reloc : [0..M] init 0; // relocated FGFR

    // FGF and FGFR bind
    [] fgf > 0 & fgfr> 0 & bnd<K
        → fgf * fgfr * k1 : (fgf'=fgf – 1) & (fgfr'=fgfr – 1) & (bnd'=bnd + 1);
    [] fgf > 0 & fgfrp> 0 & bndp<K
        → fgf * fgfrp * k1 : (fgf'=fgf – 1) & (fgfrp'=fgfrp – 1) & (bndp'=bndp + 1);
    // FGF and FGFR unbind
    [] fgf < N & fgfr< M & bnd>0
        → bnd * k2 : (fgf'=fgf +1) & (fgfr'=fgfr +1) & (bnd'=bnd – 1);
    [] fgf < N & fgfrp< M & bndp>0
        → bndp * k2 : (fgf'=fgf + 1) & (fgfrp'=fgfrp + 1) & (bndp'=bndp – 1);
    // FGFR becomes phosphorylated (FGF must be bound)
    [] bnd> 0 & bndp< K → bnd * k3 : (bnd'=bnd – 1) & (bndp'=bndp +1);
    // FGFR dephosphorylates
    [] fgfrp> 0 & fgfr< M → fgfrp * k4 : (fgfrp'=fgfrp – 1) & (fgfr'=fgfr + 1);
    [] bndp> 0 & bnd< K → bndp * k4 : (bndp'=bndp – 1) & (bnd' =bnd + 1);
    // FGFR relocates (FGFR must be phosphorylated)
    [] fgfrp> 0 & reloc< M → fgfrp * k5 : (fgfrp'=fgfrp – 1) & (reloc'=reloc + 1);
    [] bndp> 0 & reloc< M → bndp * k5 : (bndp'=bndp – 1) & (reloc'=reloc + 1);

endmodule
```

Figure 2.9 Population-based PRISM model (extending Figure 2.5)

of interaction by the number of species of each type that can take part in the reaction. As can be seen in Figure 2.9, because rates in the PRISM language can be expressions involving variables, this is straightforward to achieve with the PRISM software.

Table 2.1 lists the number of states and transitions in the individual- and population-based models as the initial number of FGF ligands and FGFR receptors ranges between 1 and 10. On the one hand, there is a rapid increase in the number of states for the individual-based approach, quickly making model checking infeasible. On the other hand, the results for the population-based model show a far more gradual increase in the state space, allowing for analysis of much larger models.

(*N*,*M*)	Individual-Based		Population-Based	
	States	Transitions	States	Transitions
(1,1)	6	9	6	9
(2,2)	52	190	21	57
(3,3)	492	3,185	56	193
(4,4)	4,816	48,804	126	509
(5,5)	47,916	699,407	252	1,140
(6,6)	480,880	9,548,278	462	2,275
(7,7)	4,849,620	125,662,133	792	4,166
(8,8)	49,045,120	1,606,974,376	1,287	7,137
(9,9)	496,798,620	20,079,205,667	2,002	11,593
(10,10)	5,036,699,152	246,135,326,874	3,003	18,029

Table 2.1 States and Transitions for Individual- and Population-Based Models

2.4.3 Model Analysis with PRISM

Two reward structures for the population model are presented in Figure 2.10. The first reward structure ("*phos*") assigns a reward to each state equal to the sum of the variables *fgfrp* and *bndp*—that is, a reward equal to the amount of FGFR that is phosphorylated. Note the difference between "*phos*" in Figure 2.10 and Figure 2.6: instead of assigning states with reward 1 or 0 depending on whether FGFR is phosphorylated, we now assign a reward equal to the sum of the two variables indicating the amount of each of the two phosphorylated forms of FGFR.

The second reward structure ("*bind*") associates a reward of 1 with all transitions of the CTMC corresponding to the binding of FGF and FGFR. In this case, we assume that we have added the label *bind* to the commands corresponding to FGF and FGFR binding (i.e., to the first and second commands of the module *POPULATION_MODEL* given in Figure 2.9). This example demonstrates that action labels can be used purely to specify reward structures, with no influence on the interaction between modules (it is, of course, important to ensure that additional unwanted interactions are not introduced in this way, such as by modules synchronizing when they should not).

We now use the temporal logic CSL and reward structures identified previously to express a number of properties of the population-based model given in Figure 2.9.

```
// amount of phosphorylated FGFR
rewards "phos"

    true : fgfrp + bndp;

endrewards
```

```
// number of bindings
rewards "bind"

    [bind] true : 1;

endrewards
```

Figure 2.10 Reward structures for the PRISM model given in Figure 2.9

- $\mathtt{P}_{=?}[\ \mathtt{true}\ \mathtt{U}\ (reloc{\geq}l_1 \wedge fgf{\leq}l_2)\ ]$; "The probability of reaching a situation where there are at least l_1 relocated FGFR receptors and the number of free FGF ligands is at most l_2."
- $(bnd{>}l_1) \Rightarrow \mathtt{P}_{\geq 0.7}[\ \mathtt{true}\ \mathtt{U}^{[0,t]}\ (reloc{\geq}l_2)\ ]$; "If the number of FGF:FGFR compounds is greater than l_1, then the probability that the amount of relocated FGFR will reach l_2 within the next t seconds is at least 0.7."
- $\mathtt{P}_{=?}[\ (reloc{=}0)\ \mathtt{U}\ (fgfp{=}l)\ ]$; "The probability that l FGFR receptors are phosphorylated before any FGFR is relocated."
- $(fgfr{=}M) \Rightarrow \mathtt{P}_{<0.05}[\ (fgfr{\geq}l_1)\ \mathtt{U}^{[t_1,t_1+t_2]}\ (fgfr{<}l_2)\ ]$; "If all FGFR receptors are free and unphosphorylated, then the probability that at least l_1 FGFR receptors remain free and unphosphorylated until time instant t_1 and that within the next t_2 seconds this number drops below l_2 is less than 0.05."
- $(fgfp+bndp{=}l) \Rightarrow \mathtt{R}\{\text{"}phos\text{"}\}_{\geq 6.8}[\ \mathtt{I}^{=t}\ ]$; "If l FGFR receptors are phosphorylated, the expected number of FGFR receptors phosphorylated t seconds later is at least 6.8."
- $\mathtt{R}\{\text{"}bind\text{"}\}_{=?}[\ \mathtt{F}\ reloc{=}l\ ]$; "The expected number of times that an FGF ligand binds with an FGFR receptor before l FGFR receptors have relocated."

As in the previous section, we give an illustration of the kind of quantitative results that can be obtained with this model. Figure 2.11(a) shows the probability of l relocations by time t for a range of different values of l when there are initially both 10 FGF ligands and 10 FGFR receptors. Figures 2.11(b) and 2.11(c) show the expected amount of phosphorylated FGFR at time t for a variety of initial configurations and two different time scales (seconds and minutes). These graphs demonstrate that after a rapid increase in phosphorylated FGFR, there is a gradual decrease as the amount of relocated FGFR increases, and that starting from a larger concentration of FGFR and FGF leads to an increase in the amount of phosphorylated FGFR.

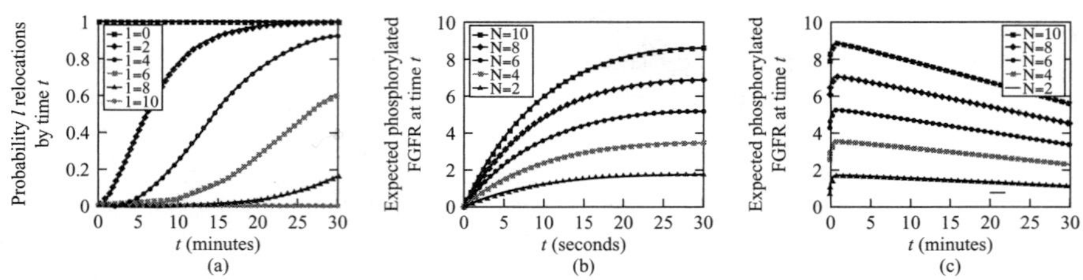

Figure 2.11 Transient properties for the population example: (a) probability of l relocations by time t; (b/c) expected phosphorylated FGFR at time t (seconds/minutes, respectively)

EXERCISES

1. Based on the model in Figure 2.9 and the reward structures in Figure 2.10, write CSL specifications for the following properties:

(a) The probability that the number of free FGF ligands does not drop below l_1 until after l_2 FGFR receptors have been relocated
(b) If the number of phosphorylated FGFR receptors is at least l, then the probability that all FGFR receptors have been relocated by time t is greater than 0.6
(c) The probability that no FGF ligands are bound to FGFR receptors until time t_1 and, before another t_2 seconds pass, a binding occurs
(d) If at most l FGFR receptors are bound, then the expected number of bindings that occur within t seconds is at least 12.4

2. Construct appropriate reward structures for properties of the model in Figure 2.9 relating to the expected time that at least l FGFR receptors are phosphorylated and the expected number of phosphorylations.
Hint: You may first need to change the PRISM model description.
Using these reward structures, write specifications for calculating the following properties:

(a) The expected time that at least l FGFR receptors are phosphorylated up until time t
(b) The expected time that at least l FGFR receptors are phosphorylated before $M-l$ receptors have been relocated
(c) The expected number of phosphorylations before a relocation occurs
(d) The expected number of phosphorylations by time t

3. Extend the model in Figure 2.9 with variables to count the number of times that an FGF ligand binds and the number of times that an FGF ligand unbinds with an FGFR receptor. Also, write CSL specifications for the properties:

(a) The probability that at most l reactions between FGF ligands and FGFR receptors (i.e., bindings or unbindings) occur within the first t seconds
(b) If no FGF ligands have bound to FGFR receptors, then the probability that l bindings occur before any unbindings occur is at least 0.5

2.5 A More Complex Model: Compartments

In this section we extend the population model of Section 2.4 to include behavior relating to the relocated FGFR. The model is motivated by the fact that FGF ligands can interact with FGFR receptors only at the cell surface; relocation causes FGFR to move inside the cell, where it loses sight of the ligand. Inside the cell, we suppose that either the FGFR is *recycled* back to the surface or it is destroyed by *degradation*. To incorporate this behavior into our model, we divide the system into the following compartments:

Main compartment: In this compartment, FGF ligands interact with FGFR receptors (i.e., at the cell surface).

Recycling compartment: This compartment receives FGFR receptors from the main compartment and at a rate of k_6 ($=1/30$ min^{-1}) reintroduces the FGFR receptors back into the main compartment.

Degradation compartment: This compartment also receives FGFR receptors from the main compartment, but in this case the proteins are degraded at a rate of k_7 ($=1/15$ min^{-1}).

In this model, we keep the assumption that the ligand FGF disappears if it is bound to FGFR when relocation occurs. In addition, we suppose that whether the proteins reach the recycling compartment versus the degradation compartment from the main compartment is probabilistic: The recycling compartment is reached with probability p and the degradation compartment is reached with probability $1-p$. Considering the rates in the CTMC representation of this model, it follows that if r is the rate in the CTMC at which FGFR receptors leave the main compartment, then the rate of FGFR being relocated to the recycling compartment is $p \cdot r$ and the rate of being relocated to the degradation compartment is $(1-p) \cdot r$.

2.5.1 The PRISM Model

Figure 2.12 shows a PRISM language model for this system based on the population model of Section 2.4 (see Figure 2.9). Each compartment is modeled by a separate module, and transitions between compartments (relocations and reintroductions) are modeled through the modules synchronizing on actions.

Figure 2.13 gives two reward structures for this model. The first reward structure ("*phos*") assigns a reward to each state equal to the total number of phosphorylated FGFR receptors in the entire system (i.e., in any of the compartments). Note that we could also have expressed this reward in a single line as follows:

$$\texttt{true} : \mathit{fgfrp}+\mathit{bndp}+\mathit{fgfrp2}+\mathit{fgfrp3}.$$

However, by taking advantage of the fact that for states/transitions that satisfy multiple guards of a reward structure the reward assigned is the sum

```
....
const double k6 = 1/(30 * 60); // half an hour to return
const double k7 = 1/(15 * 60); // time to degrade
const double p; // probability relocation causes recycling as opposed to degradation
module MAIN_COMPARTMENT
    fgf   : [0..N] init N; // free FGF
    fgfr  : [0..M] init M; // free FGFR (not phosphorylated)
    fgfrp : [0..M] init 0; // free FGFR (phosphorylated)
    bnd   : [0..M] init 0; // bound FGF:FGFR (FGFR not phosphorylated)
    bndp  : [0..M] init 0; // bound FGF:FGFR (FGFR phosphorylated)
    // FGF and FGFR bind
    [] fgf>0 & fgfr>0 & bnd<K
        → fgf * fgfr * k1 : (fgf'=fgf – 1) & (fgfr'=fgfr – 1) & (bnd'=bnd + 1);
    [] fgf>0 & fgfrp>0 & bndp<K
        → fgf * fgfrp * k1 : (fgf'=fgf – 1) & (fgfrp'=fgfrp – 1) & (bndp'=bndp + 1);
    // FGF and FGFR unbind
    [] fgf<N & fgfr<M & bnd>0
        → bnd * k2 : (fgf'=fgf + 1) & (fgfr'=fgfr + 1) & (bnd'=bnd – 1);
    [] fgf<N & fgfrp<M & bndp>0
        → bndp * k2 : (fgf'=fgf + 1) & (fgfrp'=fgfrp + 1) & (bndp'=bndp – 1);
    // FGFR becomes phosphorylated
    [] bnd>0 & bndp<K → bnd * k3 : (bnd'=bnd – 1) & (bndp'=bndp + 1);
    // FGFR becomes dephosphorylated
    [] fgfr<M & fgfrp>0 → fgfrp * k4 : (fgfr'=fgfr + 1) & (fgfrp'=fgfrp – 1);
    [] bnd<K & bndp>0 → bndp * k4 : (bnd'=bnd + 1) & (bndp'=bndp – 1);
    // phosphorylated FGFR leave main compartment
    [reloc2] bndp>0 → bndp * p * k5       : (bndp'=bndp – 1);
    [reloc3] bndp>0 → bndp * (1 – p) * k5 : (bndp'=bndp – 1);
    [reloc2] fgfrp>0 → fgfrp * p * k5       : (fgfrp'=fgfrp – 1);
    [reloc3] fgfrp>0 → fgfrp * (1 – p) * k5 : (fgfrp'=fgfrp – 1);
    // FGFR arrives from compartment 2
    [recyc] fgfr<M → (fgfr'=fgfr + 1);
    [recycp] fgfrp<M → (fgfrp'=fgfrp + 1);
endmodule
module RECYCLING_COMPARTMENT
    fgfr2  : [0..M] init 0; // free FGFR (not phosphorylated)
    fgfrp2 : [0..M] init 0; // free FGFR (phosphorylated)
    // FGFR relocates
    [reloc2] fgfrp2<M → (fgfrp2'=fgfrp2 + 1);
    // FGFR dephosphorylates
    [] fgfr2<M & fgfrp2>0 → fgfrp2 * k4 : (fgfr2'=fgfr2 + 1) & (fgfrp2'=fgfrp2 – 1);
    // FGFR returns to main compartment
    [recyc] fgfr2>0 → fgfr2 * k6 : (fgfr2'=fgfr2 – 1);
    [recycp] fgfrp2>0 → fgfrp2 * k6 : (fgfrp2'=fgfrp2 – 1);
endmodule
module DEGRADATION_COMPARTMENT
    fgfr3  : [0..M] init 0; // free FGFR (not phosphorylated)
    fgfrp3 : [0..M] init 0; // free FGFR (phosphorylated)
    deg3   : [0..M] init 0; // degraded FGFR (not phosphorylated)
    degp3  : [0..M] init 0; // degraded FGFR (phosphorylated)
    // FGFR relocates
    [reloc3] fgfrp3<M → (fgfrp3'=fgfrp3 + 1);
    // FGFR dephosphorylates
    [] fgfr3<M & fgfrp3>0 → fgfrp3 * k4 : (fgfr3'=fgfr3 + 1) & (fgfrp3'=fgfrp3 – 1);
    // FGFR degrades
    [] deg3<M & fgfr3>0 → fgfr3 * k7 : (deg3'=deg3 + 1) & (fgfr3'=fgfr3 – 1);
    [] degp3<M & fgfrp3>0 → fgfrp3 * k7 : (degp3'=degp3 + 1) & (fgfrp3'=fgfrp3 – 1);
endmodule
```

Figure 2.12 Compartments PRISM model (extending Figure 2.9)

of the rewards, often we can write the reward structure in a more readable form. The second reward structure ("*reloc*") assigns a reward of 1 to the transitions associated with FGFR receptors being relocated—that is, those labeled by *reloc2* or *reloc3*.

```
// amount of phosphorylated FGFR
rewards "phos"

    true : fgfrp + bndp;
    true : fgfrp2;
    true : fgfrp3;

endrewards
```

```
// number of relocations
rewards "reloc"

    [reloc2] true : 1;
    [reloc3] true : 1;

endrewards
```

Figure 2.13 Reward structures for the PRISM model of Figure 2.12

We now list a number of properties of the compartments model using CSL and the reward structure given in Figure 2.13:

- $\mathtt{P}_{=?}[\ \mathtt{true}\ \mathtt{U}\ (\mathit{fgfr2}+\mathit{fgfp2}\geq l)\ ]$; "The probability that at any time there are more than l FGFR receptors being recycled."
- $(\mathit{degp3}+\mathit{fgfp3}=0) \Rightarrow \mathtt{P}_{\leq 0.01}[\ (\mathit{fgfr2}+\mathit{fgfp2}=0)\ \mathtt{U}^{[0,t]}\ (\mathit{degp3}+\mathit{fgfp3}\geq l)\ ]$; "If nothing is degraded, then the probability that the number of degraded FGFR receptors reaches l before a receptor enters the recycling compartment is at most 0.01."
- $\mathtt{R}\{\text{"}\mathit{phos}\text{"}\}_{=?}[\ \mathtt{I}^{=t}\ ]$; "The expected number of FGFR receptors phosphorylated at t seconds."
- $\mathtt{R}\{\text{"}\mathit{reloc}\text{"}\}_{=?}[\ \mathtt{C}^{\leq t}\ ]$; "The expected number of relocations in the first t seconds."

Figure 2.14 shows the expected number of phosphorylated FGFR receptors in the main compartment at time t as the probability of moving to recycling when relocated varies. These results correspond to the case when initially the number of FGF ligands equals 50 and the number of FGFR receptors equals 10 (i.e., N=50 and M=10). To demonstrate the different results that are obtained through simulation as opposed to model checking, Figures 2.14(a) and 2.14(b) illustrate the results obtained with PRISM's simulator when averaging over 10 and 100 runs, respectively. Figure 2.14(c) presents the same results when using model checking. As shown in these graphs, increasing the chance of being relocated to the recycling compartment as opposed to the degradation compartment will increase the number of phosphorylated FGFR receptors in the main compartment. This outcome occurs because FGFR receptors from the main compartment relocated to the degradation compartment will not return to the main compartment, whereas those relocated to the recycling compartment will eventually return to the main compartment.

Considering the difference between the plots, we see a large fluctuation in the number of phosphorylated FGFR receptors in the main compartment when only a small number of runs are considered [Figure 2.14(a)]; as the number of runs increases, these fluctuations diminish [Figure 2.14(b)].

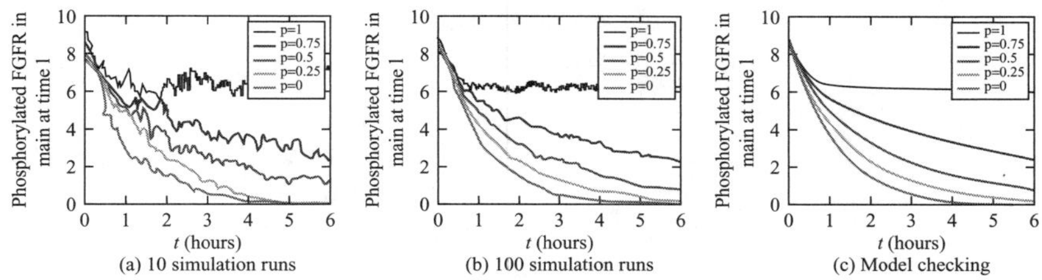

Figure 2.14 Expected amount of phosphorylated FGFR in main compartment at time t

In contrast, while employing the model-checking approach we obtain "smooth" curves [Figure 2.14(c)]. This outcome can be attributed to the fact that, when we consider an individual run, a reaction (e.g., a binding, phosphorylation, or relocation) occurs (with probability 1) at a specific time point and, therefore, its influence can be seen at that specific time point. In contrast, in model checking, an average over all possible runs is considered, and hence the probability of the reaction occurring at a certain time point is also taken into account.

EXERCISES

1. Based on the model in Figure 2.12 and the reward structures in Figure 2.13, write CSL specifications for the following properties:
   (a) The probability that eventually all FGFR receptors get degraded
   (b) If there are l_1 free FGF ligands and l_2 free FGFR receptors in the main compartment, the probability that the first degradation of an FGFR receptor occurs after time t is less than 0.1
   (c) If in the main compartment there are more than l FGF:FGFR compounds phosphorylated, then the expected number of relocations occurring by time t is at least 5.6
2. Construct an appropriate reward structure for calculating the expected number of dephosphorylations that occur in the recycling and degradation compartments of the model in Figure 2.12 and write a CSL specification for the expected number of dephosphorylations in the recycling and degradation compartments by time t.
   Hint: You will need to add extra action labels. Make sure that these items are all distinct to avoid unwanted synchronizations between modules.
3. Extend the model in Figure 2.12 with variables to count the number of receptors that are relocated to the recycling compartment and the

number of receptors that return from the recycling compartment to the main compartment. Write CSL specifications for the following properties:

(a) If l_1 receptors have entered the recycling compartment and nothing has returned to the main compartment, then, with probability at least 0.55, l_2 receptors will return within the next t seconds
(b) The probability that l_1 receptors return from the main compartment before the total number of relocations is l_2

Hint: Because nothing leaves the degradation compartment, you can use the variables in this compartment to determine the number of relocations to this compartment.

4. Rewrite the PRISM description of Figure 2.12 using a single PRISM module. Check that the new model has the same number of states and transitions as are found in the original model. In addition, using the simulation engine, generate graphs similar to those presented in Figures 2.14(a) and 2.14(b).

2.6 Related Work

In [8], PRISM has been used to study a more detailed model of the FGF signaling pathway. This model corresponds to a single instance of the pathway; that is, there can be at most one of each molecule or species. This approach has the advantage that the resulting state space is relatively small; however, the model is still highly complex due to the large number of different interactions that can occur in the pathway and is sufficiently rich to explain the roles of the components in the pathway and the means by which they interact.

In [4], PRISM is used to model the RKIP-inhibited ERK pathway using an approximate "population" approach to modeling in which concentrations are modeled by discrete abstract quantities. Also modeling the RKIP-inhibited ERK pathway, [3] demonstrates how the stochastic process algebra PEPA [9] can be used to model biological pathways. The stochastic π-calculus (pi-calculus) [18] has also been proposed as a model language for biological systems [19, 20]; this approach has so far been used in conjunction with stochastic simulation—for example, through the tools BioSpi [19] and SPiM [16]. A translation from the stochastic π-calculus to PRISM has also been developed [14].

An alternative is to use the language SBML [11], a computer-readable language based on XML for representing models of biochemical reaction networks, and the translator from SBML to the PRISM modeling language [23]. SBML is intended as a standardized representation of models that can be shared, manipulated, and analyzed using tools available in the systems biology community. Models are composed from components that permit definition of reactant species, product species, descriptions of reaction equations using MathML expressions, and the specification of kinetic laws and parameters.

The principal challenge remaining for the application of probabilistic model checking to biological systems, as in so many other domains, is the scalability of the techniques to ever larger systems and models. There is hope that some of the techniques that have already been developed in the field of formal verification, such as symmetry reduction, bisimulation minimisation, and abstraction, will prove beneficial in this area. For further details on such approaches and pointers to related work, see sources such as [8].

Acknowledgments

This work was supported in part by the EPSRC grants EP/D076625/2 and EP/D07956X/2. We would also like to thank John Heath (School of Biosciences, University of Birmingham, UK) and Oksana Tymchyshyn (School of Computer Science, University of Birmingham, UK) for their help in developing the examples.

References

[1] A. Aziz, K. Sanwal, V. Singhal, and R. Brayton. Model checking continuous time Markov chains. *ACM Transactions on Computational Logic*, 1(1): 162–170, 2000.

[2] C. Baier, B. Haverkort, H. Hermanns, and J.-P. Katoen. Model-checking algorithms for continuous-time Markov chains. *IEEE Transactions on Software Engineering*, 29(6):524–541, 2003.

[3] M. Calder, S. Gilmore, and J. Hillston. Modelling the influence of RKIP on the ERK signalling pathway using the stochastic process algebra PEPA. *Transactions on Computational Systems Biology*, 7:1–23, 2006.

[4] M. Calder, V. Vyshemirsky, D. Gilbert, and R. Orton. Analysis of signalling pathways using continuous time Markov chains. *Transactions on Computational Systems Biology*, 4:44–67, 2006.

[5] E. Clarke, E. Emerson, and A. Sistla. Automatic verification of finite-state concurrent systems using temporal logics. *ACM Transactions on Programming Languages and Systems*, 8(2):244–263, 1986.

[6] D. Gillespie. Exact stochastic simulation of coupled chemical reactions. *Journal of Physical Chemistry*, 81(25):2340–2361, 1977.

[7] H. Hansson and B. Jonsson. A logic for reasoning about time and reliability. *Formal Aspects of Computing*, 6(5):512–535, 1994.

[8] J. Heath, M. Kwiatkowska, G. Norman, D. Parker, and O. Tymchyshyn. Probabilistic model checking of complex biological pathways. *Theoretical Computer Science*, 319(3):239–257, 2008.

[9] J. Hillston. *A Compositional Approach to Performance Modelling*. Cambridge University Press, 1996.

[10] A. Hinton, M. Kwiatkowska, G. Norman, and D. Parker. PRISM: A tool for automatic verification of probabilistic systems. In H. Hermanns and J. Palsberg, editors, *Proceedings of the 12th International Conference on Tools and Algorithms for the Construction and Analysis of Systems (TACAS'06)*, volume 3920 of *Lecture Notes in Computer Science*, pages 441–444. Springer, 2006.

[11] M. Hucka, A. Finney, B. Bornstein, S. Keating, B. Shapiro, J. Matthews, B. Kovitz, M. Schilstra, A. Funahashi, J. Doyle, and H. Kitano. Evolving a lingua franca and associated software infrastructure for computational systems biology: The Systems Biology Markup Language (SBML) project. *Systems Biology*, 1(1):41–53, 2004.

[12] M. Kwiatkowska, G. Norman, and D. Parker. Probabilistic symbolic model checking with PRISM: A hybrid approach. *International Journal on Software Tools for Technology Transfer (STTT)*, 6(2):128–142, 2004.

[13] M. Kwiatkowska, G. Norman, and D. Parker. Stochastic model checking. In M. Bernardo and J. Hillston, editors, *Formal Methods for the Design of Computer, Communication and Software Systems: Performance Evaluation (SFM'07)*, volume 4486 of *Lecture Notes in Computer Science (Tutorial Volume)*, pages 220–270. Springer, 2007.

[14] G. Norman, C. Palamidessi, D. Parker, and P. Wu. Model checking probabilistic and stochastic extensions of the π-calculus. *IEEE Transactions on Software Engineering*, 2009.

[15] D. Parker. *Implementation of Symbolic Model Checking for Probabilistic Systems*. PhD thesis, University of Birmingham, Alabama, 2002.

[16] A. Phillips and L. Cardelli. Efficient, correct simulation of biological processes in the stochastic π-calculus. In M. Calder and S. Gilmore, editors, *Proceedings of the International Conference on Computational Methods in Systems Biology (CMSB'07)*, volume 4695 of *Lecture Notes in Bioinformatics*, pages 184–199. Springer–Verlag, 2007.

[17] PRISM website. `http://www.prismmodelchecker.org/`.

[18] C. Priami. Stochastic π-calculus. *Computer Journal*, 38(7):578–589, 1995.

[19] C. Priami, A. Regev, W. Silverman, and E. Shapiro. Application of a stochastic name passing calculus to representation and simulation of molecular processes. *Information Processing Letters*, 80:25–31, 2001.

[20] A. Regev, W. Silverman, and E. Shapiro. Representation and simulation of biochemical processes using the π-calculus process algebra. In R. Altman, A. Dunker, L. Hunter, and T. Klein, editors, *Pacific Symposium on Biocomputing*, volume 6, pages 459–470. World Scientific Press, 2001.

[21] W. Roscoe. *Theory and Practice of Concurrency.* Prentice-Hall, 1998.

[22] J. Rutten, M. Kwiatkowska, G. Norman, and D. Parker. *Mathematical Techniques for Analyzing Concurrent and Probabilistic Systems*, volume 23 of *CRM Monograph Series.* AMS, 2004.

[23] SBML-to-PRISM translator. `http://www.prismmodelchecker.org/sbml/`.

[24] O. Wolkenhauer, M. Ullah, W. Kolch, and K.-H. Cho. Modeling and simulation of intracellular dynamics: Choosing an appropriate framework. *IEEE Transactions on Nanobioscience*, 3:200–207, 2004.

Chapter

3

Petri Nets for Systems Biology

Monika Heiner, Robin Donaldson, and David Gilbert

Monika Heiner, Department of Computer Science, Brandenburg University of Technology, Postbox 10 13 44, 03013 Cottbus, Germany
Robin Donaldson, Department of Computing Science, 18 Lilybank Gardens, University of Glasgow, Glasgow G12 8RZ, Scotland, UK
David Gilbert, School of Information Systems, Computing and Mathematics, Brunel University, Uxbridge, Middlesex UB8 3PH, UK

Abstract This chapter introduces qualitative, stochastic, and continuous Petri nets and related analysis techniques in a rather informal way, using as a running example a model of the influence of the Raf Kinase Inhibitor Protein (RKIP) on the Extracellular signal Regulated Kinase (ERK) signaling pathway. We show how the qualitative and quantitative analyses complement each other, and how Petri nets can be used for stepwise modeling and analysis of biochemical networks as well as for structured design of systems of ordinary differential equations.

3.1 Motivation

Biochemical reaction systems have by their very nature three distinctive characteristics:

1. They are inherently bipartite; that is, they consist of two types of game players, the species and their interactions.

2. They are inherently concurrent; that is, several interactions can usually happen independently and in parallel.
3. They are inherently stochastic; that is, the timing behavior of the interactions is governed by stochastic laws.

Thus it seems to be a natural choice to model and analyze them with a formal method that shares exactly these distinctive characteristics: stochastic Petri nets.

Classical (i.e., time-free, qualitative) Petri nets combine an intuitive and executable modeling style with well-founded analysis techniques. It is for this reason that they are widely used in various application areas, where they have been proven to be useful for a qualitative verification of technical as well as "natural" systems—for example, biochemical networks such as metabolic networks, signal transduction networks, and gene regulatory networks.

However, any real system behavior happens in time. Thus the next step after qualitative analyses typically consists of quantitative analyses taking into account timing information. In the case of biochemical systems, all atomic actions (i.e., biochemical reactions) take place stochastically, which can be approximated continuously for larger quantities. Moreover, the rates of all reactions typically depend on the concentration of the involved substances. Hence systems of reaction rate equations (RREs) or ordinary differential equations (ODEs) appear well suited for quantitative modeling of biochemical networks.

In this chapter, we bridge the gap between these two worlds: the (time-free) qualitative world and the (timed) quantitative one. In doing so, we demonstrate by means of one of the standard examples used in the systems biology community—the core model of the influence of the Raf-1 Kinase Inhibitor Protein (RKIP) on the ERK signaling pathway [9]—how both sides can interact, providing different but complementary viewpoints on the same subject.

This chapter can be considered as a tutorial in stepwise modeling and analysis of biochemical networks as well as in structured design of ODEs. The qualitative model is introduced as a supplementary intermediate step, at least from the viewpoint of the biochemist accustomed to ODE modeling only, and serves mainly for model validation because this task cannot be performed on the continuous level. Having successfully validated the qualitative model, the quantitative models are derived from the qualitative one by assigning rate equations to all reactions in the network. Thus the quantitative models preserve the structure of the qualitative one, and the stochastic Petri net is simply a structured description of RREs, and the continuous Petri net is a structured description of ODEs.

The material presented here deliberately provides an informal introduction while avoiding formal notations as far as possible. See [17] for formal definitions of the technical Petri net terms used in this chapter. All concepts are explained through the running example, which is introduced in Section 3.2.

3.2 Biochemical Context

Many networks of interacting components are known to exist as part of the machinery of living organisms. Biochemical networks can be regulatory, signal transduction, or metabolic networks. Regulatory networks are used to control the ways in which genes are expressed as RNAs or proteins, whereas signal transduction networks transmit biochemical signals between or within cells. The role of metabolic networks is to synthesize essential biochemical compounds from basic components, or to degrade compounds. In this chapter we focus on signal transduction, which is the mechanism that enables a cell to sense changes in its environment and to make appropriate responses. The basis of this mechanism is the conversion of one kind of signal into another.

The ERK pathway (also called Ras/Raf, or the Raf-1/MEK/ERK pathway) is a ubiquitous pathway that conveys cell division and differentiation signals from the cell membrane to the nucleus. Ras is activated by an external stimulus, via one of many growth factor receptors; it then binds to and activates Raf-1 to become Raf-1* (activated Raf), which in turn activates MAPK/ERK Kinase (MEK), which in turn activates Extracellular signal Regulated Kinase (ERK). This cascade (Raf-1 → Raf-1* → MEK → ERK) of protein interaction controls cell differentiation, with the ultimate effect depending on the activity of ERK. An important area of experimental scientific investigation is the role that the Raf-1 Kinase Inhibitor Protein (RKIP) plays in the behavior of this pathway. The hypothesis is

Number	Precursors			Products		
*r*1	Raf-1*	+ RKIP	→	Raf-1*_RKIP		
*r*2	Raf-1*_RKIP		→	Raf-1*	+ RKIP	
*r*3	Raf-1*_RKIP	+ ERK-PP	→	Raf-1*_RKIP_ERK-PP		
*r*4	Raf-1*_RKIP_ERK-PP		→	Raf-1*_RKIP	+ ERK_PP	
*r*5	Raf-1*_RKIP_ERK-PP		→	Raf-1*	+ ERK	+ RKIP-P
*r*6	MEK-PP	+ ERK	→	MEK-PP_ERK		
*r*7	MEK-PP_ERK		→	MEK-PP	+ ERK	
*r*8	MEK-PP_ERK		→	MEK-PP	+ ERK-PP	
*r*9	RKIP-P	+ RP	→	RKIP-PP_RP		
*r*10	RKIP-P_RP		→	RKIP-P	+ RP	
*r*11	RKIP-P_RP		→	RKIP	+ RP	

Note: Raf-1* is often written as Raf-1Star (if required by the tools used).

Table 3.1 The Reaction Equations of the Running Example

that RKIP can inhibit activation of Raf-1 by binding to it and thereby disrupting the interaction between Raf-1 and MEK; in this way, it plays a part in regulating the activity of the ERK pathway. The subsystem we consider here, taken from [9], comprises the 11 reactions summarized in Table 3.1.

3.3 Qualitative Approach

In this section we introduce place/transition Petri nets, and interpret them in the standard way to model and analyze the pathway of interest.

3.3.1 Qualitative Modeling

Let us start with a bit of history. Petri nets, as we understand them today, have been initiated by concepts proposed by Carl Adam Petri in his Ph.D. thesis in 1962 [30]. The first substantial results making up the still-growing body of Petri net theory appeared around 1970. The initial textbooks devoted to Petri nets were issued in the beginning of the 1980s. Since that time, Petri nets have been employed for technical and administrative systems in numerous application domains. Their employment in systems biology was first published in three sources [21, 32, 33]. In the meantime, the three biochemical network types as well as combinations of them have been investigated with different kinds of Petri nets—qualitative as well as quantitative ones.

The idea to represent biochemical networks by Petri nets is rather intuitive and has been mentioned by Carl Adam Petri himself in one of his internal research reports on interpretation of net theory in the 1970s. Place/transition Petri nets (or just "nets," for short) are weighted, directed, bipartite graphs with four basic ingredients (compare with Figure 3.1).

First, there are two types of nodes: *places*, described as $S = \{s_1, \ldots, s_m\}$ and in the figures represented by circles, and *transitions*, described as $R = \{r_1, \ldots, r_n\}$ and in the figures represented by rectangles. Places usually model passive system components such as conditions, species, or any kind of chemical compounds (e.g., proteins or proteins complexes) playing the

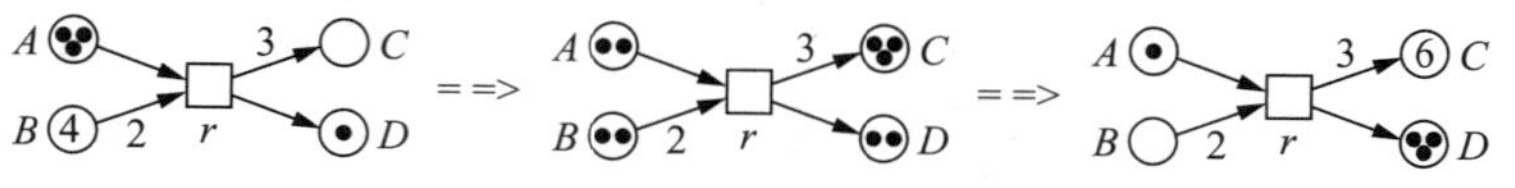

Figure 3.1 The Petri net for the single reaction system $A + 2 \cdot B \rightarrow 3 \cdot C + D$ and three of its markings (states), each connected by a firing of the transition r. The transition is no longer enabled in the marking reached after these two single firing steps.

role of precursors or products. Transitions usually stand for active system components such as atomic actions or any kind of chemical reactions (e.g., association, disassociation, phosphorylation, or dephosphorylation).

Second, the directed arcs, represented as arrows, (only) connect nodes of different types. They go from precursors to reactions (ingoing arcs), and from reactions to products (outgoing arcs). Enzymes establish side conditions and are connected in both directions with the reaction they catalyze. We define four types of sets:

- The preplaces of a transition, consisting of the reaction's precursors
- The postplaces of a transition, consisting of the reaction's products
- The pretransitions of a place, consisting of all reactions, producing this species
- The posttransitions of a place, consisting of all reactions, consuming this species

Third, arcs are weighted by non-negative integers (natural numbers), whereby the arc weight may be read as the multiplicity of the arc, reflecting known stoichiometries. The arc weight 1 is the default value and is usually not given explicitly.

Fourth, a place carries an arbitrary number of *tokens*, represented as black dots or a non-negative integer. The number 0 is the default value; it is usually not given explicitly. Tokens can be interpreted as the available amount of a given species in number of molecules or moles, or in any abstract (i.e., the discrete concentration level).

In the most abstract way, a concentration can be thought of as being "high" or "low" (present or absent, respectively); we get two levels. Generalizing this Boolean approach, any continuous concentration range can be divided into a finite number of equally sized subranges (equivalence classes), so that the concentrations within the ranges can be considered to be equivalent. The current number of tokens on a place will then specify the current level of the species' concentration—for example, the absence of tokens specifies level 0. In the following discussion, when speaking in terms of level semantics, we always give the highest level number.

A place is called marked, if it has at least one token; otherwise, it is called clean (unmarked). The tokens on all places constitute the *marking* of the net, which represents the current state of the system.

Once the net structure is understood, the next step is to bring the net to life by moving the tokens through the net. We need to introduce the firing rule for the token game, which consists of two parts: the precondition and the firing itself.

- A transition is *enabled* if all its preplaces carry at least as many tokens as required by the weights of the corresponding ingoing arcs.

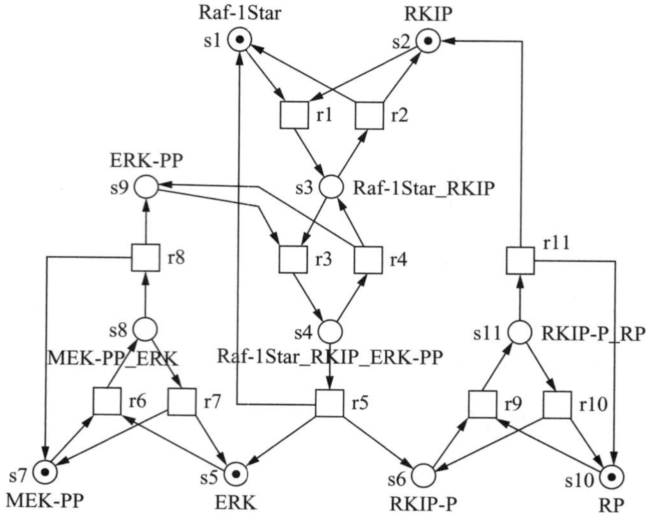

PUR	ORD	HOM	NBM	CSV	SCF	CON	SC	FTO	TFO	FPO	PFO	NC
Y	Y	Y	Y	N	N	Y	Y	N	N	N	N	ES
DTP	CPI	CTI	SCTI	SB	k-B	1-B	DCF	DSt	DTr	LIV	REV	
Y	Y	Y	N	Y	Y	Y	N	0	N	Y	Y	

Figure 3.2 The Petri net for the core model of the RKIP pathway, consisting of 11 places and 11 transitions (by chance). The places {s1, ..., s11} stand for proteins or protein complexes. Complexes are indicated by an underscore "_" between the protein names, and phosphorylated forms by the suffix -P or -PP. The transitions {r1, ..., r11} model the reactions. The preplaces of a transition correspond to the reaction's precursors, and its postplaces to the reaction's products. The layout follows the suggestions by the graphical notation used in [9]. The initial marking is constructed systematically using standard Petri net analysis techniques. At the bottom, the two-line result vector as produced by Charlie [8] is given. Properties of interest for this biochemical network in this vector are explained in the text.

- An enabled transition *may* fire; that is, a transition is never *forced* to fire. The firing of a transition removes from all its preplaces as many tokens as specified by the ingoing arc weights, and adds to all its postplaces as many tokens as specified by the outgoing arc weights. The firing happens atomically and does not consume any time.

Figuratively, the firing of a transition moves tokens from its preplaces to its postplaces, while possibly changing the number of tokens. Generally, the firing of a transition changes the formerly current marking to a new reachable one, where some transitions are no longer enabled, while others become newly enabled. The repeated firing of transitions establishes the behavior of the net. The whole net behavior consists of all possible partially ordered firing sequences (partial-order semantics) or all possible totally ordered firing sequences (interleaving semantics), respectively, and the *state space* of the net of all reachable markings. In this introduction we confine ourselves to the interleaving semantics.

In this modeling spirit, we create a place/transition Petri net of our running example (see Figure 3.2), given in [9] in the style of a bipartite graph. Circles (places) stand for the (local) states of a protein or protein complex and are labeled with the corresponding name; complexes are indicated by an underscore "_" between the protein names. For example, Raf-1Star and RKIP are proteins, and Raf-1Star_RKIP is a protein complex formed from Raf-1Star and RKIP. Suffixes -P and -PP denote a single and a double phosphorylated protein, respectively—for example, RKIP-P and ERK-PP. If appropriate, we will use the shortcuts s1, s2, ..., instead of the full names. There are 11 species, each of which is associated with a discrete concentration. In the most abstract way, these concentrations can be thought of as being "high" or "low" (present or absent, respectively). Later we will refine this Boolean approach to several discrete concentration levels.

Rectangles (transitions) stand for reactions, with reversible reactions being modeled by two opposite reactions. In this pathway, reactions comprise protein complexation and decomplexation events, often accompanied by phosphorylation or dephosphorylation. For example, Raf-1Star and RKIP combine in a forward reaction to form Raf-1Star_RKIP, which can disassociate in a backward reaction into Raf-1Star and RKIP, or which can combine with ERK-PP to form the complex Raf1Star_RKIP_ERK-PP. Reactions are labeled with r1, r2, ..., to ease referencing. In the case of reversible reactions, the labels are given by rn, rn+1, denoting that rn is the forward reaction and rn+1 the backward reaction. For example, the forward reaction for the combination of Raf-1Star and RKIP is r1, and the disassociation of Raf-1Star_RKIP is r2.

Figure 3.3 shows examples of the basic structures in Petri net notation, which could help in designing your own Petri net.

3.3.2 Qualitative Analysis

A preliminary step will usually execute the net, which allows us to experience the model behavior by following the token flow.[1] Having established initial confidence in the model by playing the token game, the system needs to be formally analyzed. Formal analyses are exhaustive, unlike the token game, which exemplifies the net behavior.

---

[1] If the reader would like to give it a try, just download our Petri net tool [38].

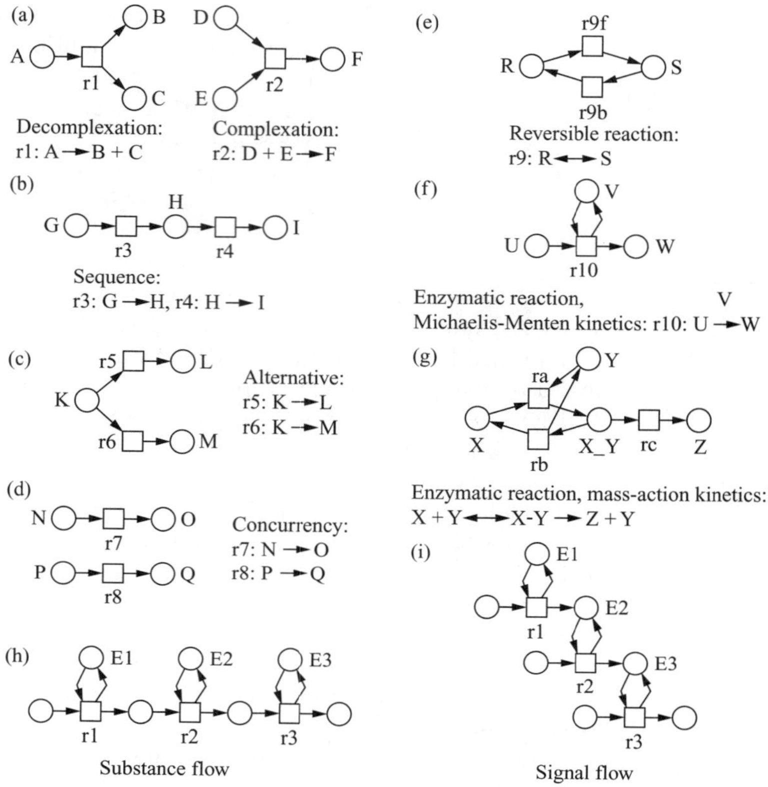

Figure 3.3 The Petri net components for some typical basic structures. **(a)** decomplexation and complexation; **(b)**–**(d)** the three basic structural principles: sequence, alternative, and concurrency; **(e)**–**(g)** reversible reaction, enzymatic reaction in the Michaelis-Menten approach, and enzymatic reaction in the mass-action approach; **(h)** the essential distinctive principles of metabolic networks (substance flow) and **(i)** signal transduction networks (signal flow).

The Petri net enjoys all of the pleasant general properties a Petri net insider could dream of, including three orthogonal basic behavioral net properties:

- *Boundedness.* For every place it holds that, whatever happens, the maximal number of tokens on this place is bounded by a constant. This precludes overflow by an unlimited increase of tokens.
- *Liveness.* For every transition it holds that, whatever happens, it will always be possible to reach a state where this transition gets enabled. In a live net, all transitions are able to contribute to the net behavior

forever, which precludes dead states (i.e., states where none of the transitions is enabled).

- *Reversibility.* For every state it holds that, whatever happens, the net will always be able to reach this state again. Thus the net has the capability of self-reinitialization.

In most cases these are requirable properties. The decision about the first two properties can be made for our example in a static way (i.e., without constructing the state space), while the last property requires dynamic analysis techniques (i.e., the construction of the state space). The essential steps of the systematic analysis procedure for our running example are given in more detail next. They represent a typical pattern of how to proceed, so they may be taken as a recipe for how to analyze your own system.

1. *Structural properties.* The following three structural properties reflect the modeling approach and can be taken as preliminary consistency checks to preclude production faults in drawing the net.

The net is *pure* (PUR); that is, there are no place and transition connected in both directions. Thus the net structure is fully represented by the incidence matrix, which is used for the calculation of the P- and T-invariants, as shown in step 2.

The net is *ordinary* (ORD); that is, all arc weights are equal to 1. This includes homogeneity (HOM); that is, the outgoing arcs of each place have the same multiplicity, which is a necessary prerequisite for the Deadlock Trap Property (DTP), as discussed in step 4.

The net is *strongly connected* (SC); that is, there is a directed path between all pairs of nodes. This precludes boundary nodes, which exist in four types:

- **Input transition:** a transition without preplace (FT0)
- **Output transition:** a transition without postplace (TF0)
- **Input place:** a place without pretransition (FP0)
- **Output place:** a place without posttransition (PF0)

Boundary nodes model interconnections of an open system with its environment. However, the given net is self-contained—that is, a closed system. Therefore, to make the net live, we have to construct an initial marking, as in step 3.

Finally, there are transitions sharing a preplace; for example, the transitions r2 and r3 share the preplace Raf-1Star_RKIP. Consequently, these transitions compete for the tokens on the shared preplace. Such a situation is called *structural conflict.* If each transition is involved in one conflict at most, the net belongs to the *extended simple* (ES) structural class, which is the case for our running example. Hence, we know that the net has the ability to be live independent of time: If it is live, then it remains live under any timing. Remarkably, neglecting the backward directions of all reversible reactions removes all structural conflicts in the net structure, which makes this reduced net into a *synchronization graph* (SG). This structural net class is characterized by being free of structural conflicts (SCF).

2. *Static decision of marking-independent behavioral properties.* A beneficial technique in model validation is to check all invariants for their biological plausibility. We will cover the essential technical terms shortly.

At the beginning, we need to represent the net structure by a matrix, which opens the door to analysis techniques based on linear algebra. The *incidence matrix* of a Petri net is an integer matrix ₵ with a row for each place and a column for each transition. A matrix entry ₵ (s, r) gives the token change on place s by the firing of transition r. Thus a preplace of r, which is not a postplace of r, has a negative entry, while a postplace of r, which is not a preplace of r, has a positive entry. Each of these entries corresponds to the arc multiplicities. The entry for a place, which is a preplace as well as a postplace of a transition, gives the difference of the multiplicities of the transition's outgoing arc minus the transition's ingoing arc. In this case we lose information; the non-ordinary net structure cannot be reconstructed uniquely out of the incidence matrix.

The columns of ₵ are place vectors—that is, vectors with as many entries as there are places. These vectors describe the token change on a marking by the firing of the transition defining the column index. The rows of ₵ are transition vectors—that is, vectors with as many entries as there are transitions. These vectors describe the influence of all transitions on the tokens in the place, defining the row index. For stoichiometric reaction networks (e.g., metabolic networks), the incidence matrix coincides with the stoichiometric matrix.

A non-zero and non-negative integer place vector x is called a *P-invariant* if $x \cdot$₵ $= 0$. In words, for each transition, it holds that multiplying the P-invariant with the transition's column vector yields zero.

Thus the total effect of each transition on the P-invariant is zero, which explains its interpretation as a token conservation component. A P-invariant x stands for a set of places over which the weighted sum of tokens is constant and independent of any firing. Thus, for any markings m_1, m_2 that are reachable by the firing of transitions, it holds that $x \cdot m_1 = x \cdot m_2$. In the context of metabolic networks, P-invariants reflect substrate conservations; in signal transduction networks, P-invariants often correspond to the several states of a given species (protein or protein complex). A place belonging to a P-invariant is obviously bounded.

Likewise, a non-zero and non-negative integer transition vector y is called *T-invariant* if ₵ $\cdot y = 0$. In words, for each place, it holds that multiplying the place's row with the T-invariant yields zero. Thus the total effect of the T-invariant on a marking is zero. A T-invariant has two interpretations in the given biochemical context.

- The entries of a T-invariant represent a multiset of transitions, which by their partially ordered firing reproduce a given marking (i.e., basically occurring one after the other). This partial-order sequence of the T-invariant's transitions may contribute to a deeper understanding of the net behavior. A T-invariant is called *feasible* if such a behavior is actually possible in the given marking situation.
- The entries of a T-invariant may also be read as the relative firing rates of transitions, all of them occurring permanently and concurrently. This activity level corresponds to the steady-state behavior.

The two transitions modeling the two directions of a reversible reaction always create a minimal T-invariant; thus they are called *trivial T-invariants.*

The computation of all P-invariants (T-invariants) requires solving the homogeneous linear equation system $x \cdot ₵ = 0$ ($₵ \cdot y = 0$) over the non-negative integers. If solutions exist, then there are infinitely many. To characterize the solution space, we need a generating system. For this purpose we introduce the next terms, which hold equally for P- and T-invariants.

The set of nodes corresponding to an invariant's non-zero entries constitutes its *support.* An invariant x is called *minimal* if its support does not contain the support of any other invariant, and if the greatest common divisor of all non-zero entries of x is 1. The set of all minimal invariants of a given net is unique and is a generating system for all invariants. To compute all invariants out of the minimal ones, we use the following operations: multiplication with a non-negative integer, addition of invariants, and division by a common divisor.

A minimal P-invariant (T-invariant) defines a connected subnet, consisting of its support, its pretransitions and posttransitions (preplaces and postplaces), and all arcs in between. These minimal self-contained subnets may be read as a decomposition into token preserving or state-repeating modules, which should have an enclosed biological meaning. Minimal invariants generally overlap, and in the worst-case scenario there are exponentially many of them.

A net is covered by P-invariants, shortly CPI (covered by T-invariants, shortly CTI), if every place (transition) belongs to a P-invariant (T-invariant). CPI causes structural boundedness (SB)—that is, boundedness for any initial marking. CTI is a necessary condition for bounded nets to be live.

The net under consideration has five minimal P-invariants covering the net (CPI); consequently, the net is structurally bounded (SB). All of the

P-invariants x_i contain only entries of 0 and 1, which allows for the use of a shorthand specification by just giving the names of the places involved:

x_1 = (**Raf-1Star**, Raf-1Star_RKIP, Raf-1Star_RKIP_ERK-PP)
x_2 = (**MEK-PP**, MEK-PP_ERK)
x_3 = (**RP**, RKIP-P_RP)
x_4 = (**ERK**, ERK-PP, MEK-PP_ERK, Raf-1Star_RKIP_ERK-PP)
x_5 = (**RKIP**, Raf-1Star_RKIP, Raf-1Star_RKIP_ERK-PP, RKIP-P_RP, RKIP-P)

Here each P-invariant x_i stands for a reasonable conservation rule, where the species preserved given in bold is the first name in the invariant. Due to the chosen naming convention, this particular name also appears in all the other place names of the same P-invariant.

The net under consideration is also covered by T-invariants (CTI). Besides the expected four trivial T-invariants for the four reversible reactions, there is only one nontrivial minimal T-invariant y = (r1, r3, r5, r6, r8, r9, r11), comprising the whole net, but without the backward reactions. The subnet defined by this T-invariant describes the essential partial-order behavior of the modeled system.

The automatic identification of nontrivial minimal T-invariants is in general useful as a method to highlight important parts of a network, and hence aid its comprehension by biochemists, especially when the entire network is too complex to easily comprehend.

Notably, our running example collapses into a synchronization graph, if we remove the backward directions of the reversible reactions. In a strongly connected ordinary synchronization graph, each minimal cycle corresponds to a minimal P-invariant, and the set of all transitions constitutes a T-invariant. This confirms from a different perspective the invariant's results, which we have just discussed.

All of the properties mentioned earlier relate only to the structure; in other words, they are valid independently of the initial marking. To proceed we first need to generate an initial marking.

3. *Initial marking construction.* For a systematic construction of the initial marking, we consider the following criteria:

- Each P-invariant needs at least one token.
- All (nontrivial) T-invariants should be feasible, meaning that the transitions making up the T-invariant's multiset can actually fire in an appropriate order.
- Additionally, it is common sense to look for a minimal marking (as few tokens as possible), which guarantees the required behavior.
- Within a P-invariant, choose the species with the most *inactive* or the *monomeric* state.

Taking all these criteria together, the initial marking on hand is that Raf-1Star, RKIP, ERK, MEK-PP, and RP each get one token, while all remaining places are clean. With this initial marking, the net is covered by 1-P-invariants (exactly one token in each P-invariant), so the net is 1-bounded (indicated as 1-B in the analysis result vector; compare with Figure 3.2). That result is in perfect accordance with the understanding that in signal transduction networks a P-invariant comprises all the different states of one species. Obviously, each species can be in only one state at any time.

This initial marking differs from the initial concentrations used in [9] as part of the method to estimate rate parameters required for an ODE model of the RKIP pathway. We will later see that this initial marking is equivalent to the initial marking that we have constructed, and that, in fact, both markings are members of a larger equivalence class of markings.

With the chosen marking, we can check the nontrivial minimal T-invariant (see step 2) for realizability, which then involves the realizability of the trivial T-invariants. We obtain an infinite partial-order run, the beginning of which is given as labeled condition/event net in Figure 3.4. We get this run by unfolding the behavior of the subnet induced by the T-invariant, whereby any concurrency is preserved. Here, transitions represent events, labeled by the name of the reaction taking place, while places stand for binary conditions, labeled by the name of the species, that are set or reset by the event, respectively. This run can be characterized in a shorthand notation by the following set of partially ordered words out of the alphabet of all transition labels R:

$$\{ \text{ (r1, (r6; r8)); [(r3; r5);((r9; r11; r1), (r6; r8))]* } \}$$

where ";" stands for "sequentiality" and "," for "concurrency". This partial-order behavior gives further insight into the dynamic behavior of the network, which may not be apparent from the standard net representation. In our running example, the partial-order representation illustrates the central role of Raf-1Star and the fact that certain reactions take place sequentially, other reactions take place independently, and some reactions are constrained by the need for the presence of two or more precursors produced from prior reactions. For example, the reaction r1 requires the participation of both proteins Raf-1* and RKIP, which are produced by the reactions r5 and r11, respectively.

Having established and justified our initial marking we proceed to the next steps of the analysis.

4. *Static decision of marking-dependent behavioral properties.* The following advanced structural Petri net properties can be decided by combinatorial algorithms. First, we need to introduce two new notions.

A nonempty set of places $D \subseteq S$ is called a *structural deadlock* if every transition that fires tokens onto a place in this structural deadlock set also

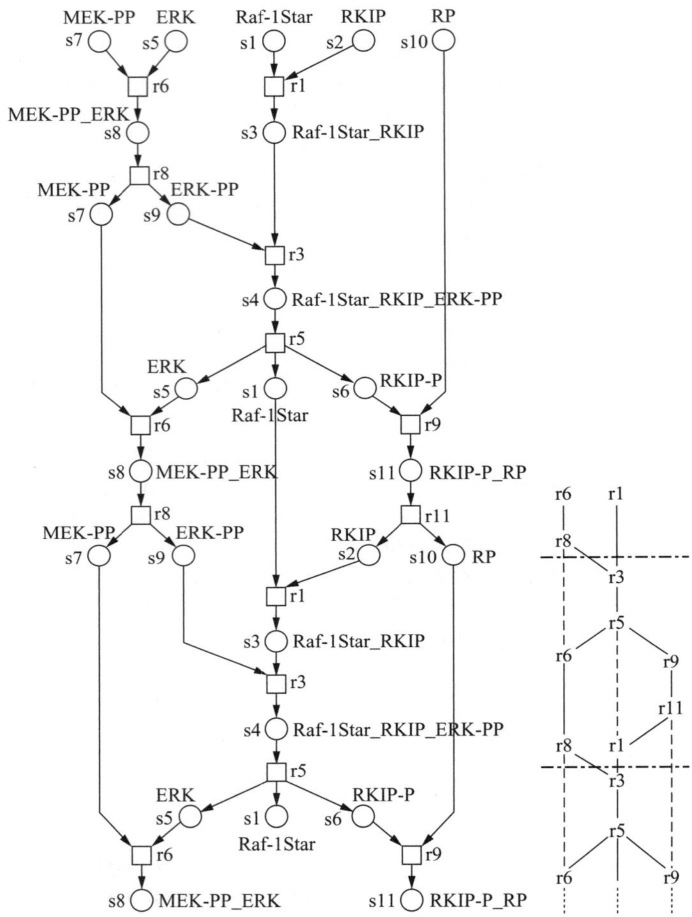

Figure 3.4 The beginning of the infinite partial-order run of the nontrivial minimal T-invariant $y =$ (r1, r3, r5, r6, r8, r9, r11) of the place/transition Petri net given in Figure 3.2. We obtain this run by unfolding the behavior of the subnet induced by the T-invariant, whereby any concurrency is preserved. Here, transitions represent events, labeled by the name of the reaction taking place, while places stand for binary conditions, labeled by the name of the species, that are set or reset by the event, respectively. On the right, a shorthand notation is given, showing the cuts between two "rounds."

has a preplace in this set—in other words, if the set of pretransitions of D is contained in the set of posttransitions of D. Pretransitions of a structural deadlock cannot fire if the place set is clean. Therefore, a structural deadlock cannot get tokens again as soon as it is clean, and then all its posttransitions are dead.

A nonempty set of places $Q \subseteq S$ is called a *trap* if every transition that subtracts tokens from a place of the trap set also has a postplace in this set—in other words, if the set of posttransitions of Q is contained in the set of pretransitions of Q. Posttransitions of a trap always return tokens to the place set. Therefore, once a trap contains tokens, it cannot become clean again.

Structural deadlock and trap are closely related, but contrasting, notions. When they come on their own, we get usually deficient behavior. However, both notions have the power to complement each other perfectly. A Petri net has the *Deadlock-Trap Property* (DTP), if every structural deadlock contains a marked trap. The DTP can still be decided by structural reasoning only. The importance becomes clear by the following theorem: An ordinary extended simple net having the DTP is live.

Our running example is extended simple and the DTP holds; therefore the net is live (LIV). As a consequence, we are in the fortunate position of being able to decide liveness by structural reasoning only; the state space does not need to be constructed for this decision.

5. *Dynamic decision of behavioral properties.* To decide reversibility (REV), we need to construct the state space. This could be done according to the partial-order semantics or the interleaving semantics. To keep things simple we consider here the reachability graph only. The nodes of a reachability graph represent all possible states (markings) of the net. The arcs in between are labeled by single transitions, the firing of which causes the related state change. Altogether, the reachability graph gives us a finite automaton representation of all possible single-step firing sequences. Consequently, concurrent behavior is described by enumerating all interleaving firing sequences, so it reflects the behavior of the net in the interleaving semantics.

Because we already know that the net under consideration is bounded, we also know that the reachability graph has to be finite. Here, the reachability graph has 13 states (out of $2048 = 2^{11}$ theoretically possible ones, due to the given 1-boundedness); see Figure 3.5 and Table 3.3. Generally, reachability graphs tend to be huge. Table 3.2 shows the size of the reachability graph with a different number of concentration levels.

Number of Levels	Number of States	Number of Levels	Number of States	Number of Levels	Number of States
1	13	15	368,220	40	79,414,335
5	1,974	20	1,696,618	50	2.834 e+08
7	8,646	25	5,723,991	100	1.591 e+13
9	28,171	30	15,721,464	250	3.582 e+12
10	47,047	35	37,314,537	500	2.231 e+14

Table 3.2 The Number of States in the State Space as the Number of Levels in the Model Increases.

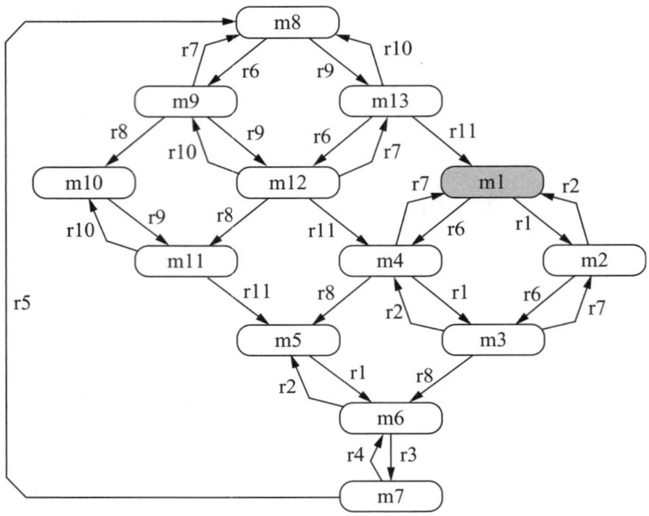

Figure 3.5 The reachability graph of the place/transition Petri net given in Figure 3.2. Nodes represent states (markings), while the arcs are labeled with the transition responsible for a given change of states. For a complete list of the state descriptions, see Table 3.3. The initial state is highlighted in gray. In the reachability graph, the two concurrent transition sequences (r6; r8) and (r9; r11; r1) compare the partial-order run in Figure 3.4, going from m8 to m5, are represented by all possible totally ordered (interleaving) sequences. Thus the reachability graph reflects the interleaving semantics.

As is clear in Table 3.2, as the number of levels increases, the number of states in the reachability graph grows extremely rapidly. The term *state explosion* has been coined for this phenomenon. Independent of the size, the reachability graph we get is strongly connected. Therefore, the Petri net is reversible: The initial system state is always reachable again, and the system has the capability of self-reinitialization. Further, the liveness of the net has already been decided structurally, so we know that each transition (reaction) appears at least once as an arc label in this reachability graph.

Moreover, from the viewpoint of the qualitative model, all of these 13 states are equivalent; any of those 13 states could be taken as initial state resulting in exactly the same total (discrete) system behavior (likewise for the larger state sets). That finding is in perfect accordance with the observations gained during quantitative analyses, as discussed in the next section.

This concludes the analysis of *general* behavioral net properties or those properties we can speak about in syntactic terms only, without any semantic knowledge. The next step takes a closer look at *special* behavioral net properties, reflecting the expected special functionality of the network.

Number	Species	m1	m2	m3	m4	m5	m6	m7	m8	m9	m10	m11	m12	m13
s1	Raf-1*	1	0	0	1	1	0	0	1	1	1	1	1	1
s2	RKIP	1	0	0	1	1	0	0	0	0	0	0	0	0
s3	Raf-1*_RKIP	0	1	1	0	0	1	0	0	0	0	0	0	0
s4	Raf-1*_RKIP_ERK-PP	0	0	0	0	0	0	1	0	0	0	0	0	0
s5	ERK	1	1	0	0	0	0	0	1	0	0	0	0	1
s6	RKIP-P	0	0	0	0	0	0	0	1	1	1	0	0	0
s7	MEK-PP	1	1	0	0	1	1	1	1	0	1	1	0	1
s8	MEK-PP_ERK	0	0	1	1	0	0	0	0	1	0	0	1	0
s9	ERK-PP	0	0	0	0	1	1	0	0	0	1	1	0	0
s10	RP	1	1	1	1	1	1	1	1	1	1	0	0	0
s11	RKIP-P_RP	0	0	0	0	0	0	0	0	0	0	1	1	1

Table 3.3 States of the Reachability graph Given in Figure 3.5. These are used as the initial 13 "good" state configurations in the quantitative analyses.

6. *Model checking of special behavioral properties.* To validate the model, it is often of interest to prove—besides the general properties—additional special properties that reflect the intended functionality of the network. We have to specify these special properties in an unambiguous language. Temporal logics, a mathematical mechanism, has been proven to be best suited for this purpose. It provides a flexible formalism that considers the validity of propositions in relation to the execution of the model.

One of the widely used temporal logics is *Computational Tree Logic* (CTL). Its name reflects the data structure used—the computational tree, which we get by unwinding the reachability graph. Thus CTL represents a branching time logic with interleaving semantics. The analysis technique, which decides whether a temporal-logic property holds in a model, is called *model checking*, and the tools implementing the algorithms are called *model checkers*. Model checking generally requires boundedness. If the net is 1-bounded, there exists a particularly rich choice of model checkers [35] that achieve their efficiency by exploiting sophisticated data structures and algorithms. In the case of our rather simple example, the variety of model checkers is not important, and the properties could even be checked manually. However, as mentioned previously, the state space of more complex networks is subject to the famous state explosion problem and, therefore, calls for automatic evaluation.

The application of this analysis approach requires an understanding of temporal logics. CTL, like any temporal logic, is an extension of a classical (propositional) logic. The atomic propositions consist of statements on the current token situation in a given place. In the case of 1-bounded models, places can be read as Boolean variables, which allows for propositions such as RKIP-P instead of $m(\text{RKIP-P}) = 1$, where $m(s)$ yields the number of tokens on s. Likewise, places are read as integer variables for bounded models.

Atomic propositions can be combined to composed propositions using the standard logical operators: $\neg$ (negation), $\wedge$ (conjunction), $\vee$ (disjunction), and $\rightarrow$ (implication). For example, a disjunction expression would have the form RKIP-P $\vee$ RKIP-P_RP.

The truth value of a proposition may change by the execution of the net. For example, the proposition RKIP does hold in state m1, but not in state m2. Such temporal relations between propositions are expressed by still other temporal operators. In CTL, there are basically four of these additional operators (ne**X**t, **F**inally, **G**lobally, **U**ntil), which come in two versions (**E** for Existence, **A** for All), for a total of eight operators. Let $\phi_{[1,2]}$ be an arbitrary temporal-logic formulae. Then the following formulae hold in state m,

- **EX** ϕ : if there is a state reachable by one step where ϕ holds.
- **EF** ϕ : if there is a path where ϕ holds finally (i.e., at some point).
- **EG** ϕ : if there is a path where ϕ holds globally (i.e., forever).
- **E** (ϕ_1 **U** ϕ_2) : if there is a path where ϕ_1 holds until ϕ_2 holds.

The other four operators, which we obtain by replacing the **E**xistence operator with the **A**ll operator, are defined likewise by extending the requirement "there is a path" to "for all paths." A formula holds in a net, if it holds in the net's initial state.

We instantiate some of the generic property patterns provided in [6] and get the following samples of meaningful statements for our running example, whose truth can be determined via model checking. When examining these examples, remember that places are interpreted as Boolean variables in the formulae, so as to simplify notation.

- **Property Q1:** There is a reachable state where ERK is phosphorylated and RKIP is not phosphorylated.
  EF [(ERK-PP $\vee$ Raf-1Star_RKIP_ERK-PP) $\wedge$
  (RKIP $\vee$ Raf-1Star_RKIP $\vee$ Raf-1Star_RKIP_ERK-PP)]
  The given state specification applies to state m5, and this state is, of course, reachable from m1; the shortest firing sequence is r6; r8 (compare Figure 3.5).
- **Property Q2:** The phosphorylation of ERK (to ERK-PP) is independent of the phosphorylated state of RKIP.
  AG [(ERK $\wedge$ $\neg$(RKIP-P $\vee$ RKIP-P_RP)) $\rightarrow$
  E [$\neg$(RKIP-P $\vee$ RKIP-P_RP) **U** ERK-PP]]
  In words, forever it holds (AG), if we are in a state where ERK is marked, but neither RKIP-P nor RKIP-P_RP; that is, if RKIP is not phosphorylated, then there is a path (E), along which in all states RKIP is not phosphorylated until (U) we reach a state were ERK-PP is marked.

- **Property Q3:** A cyclic behavior w.r.t. the presence/absence of RKIP is possible forever.

  $\mathbf{AG}\ [(\text{RKIP} \rightarrow \mathbf{EF}\ (\neg\text{RKIP})) \wedge (\neg\text{RKIP} \rightarrow \mathbf{EF}\ (\text{RKIP}))]$

  In words, forever it holds (AG), if we are in a state where RKIP is marked, then there is a path to a state (EF), where RKIP is not marked anymore, and vice versa.

  Taking into consideration that the backward reactions are comparatively slow (see Table 3.4), which allows us to neglect them, we are able to expect a much stronger property: The cyclic behavior w.r.t. the presence/absence of RKIP is not just possible forever, it is even guaranteed.

  $\mathbf{AG}\ [(\text{RKIP} \rightarrow \mathbf{AF}\ (\neg\text{RKIP})) \wedge (\neg\text{RKIP} \rightarrow \mathbf{AF}\ (\text{RKIP}))]$

  In words, forever it holds (AG), if we are in a state where RKIP is marked, then on all paths we reach finally a state (AF) where RKIP is not marked anymore, and vice versa. This stronger property does not hold in the full net (with the backward reactions), because the reversible reactions establish infinite cycles (paths) that prevent the system from reaching a state as required.

Temporal logic is an extremely powerful and flexible language to describe special properties, but the user needs some experience to get accustomed to it. Applying this analysis technique requires seasoned understanding of the network under investigation combined with the skill to correctly express the expected behavior in temporal logics. Section 3.4 explores ways to employ the same technique in a quantitative setting.

3.3.3 Summary

To summarize the preceding validation steps, the model has passed the following validation criteria:

- **Validation criterion 0.** All expected structural properties hold, and all expected general behavioral properties hold.
- **Validation criterion 1.** The net is CPI, and there are no minimal P-invariants without biological interpretation.
- **Validation criterion 2.** The net is CTI, and there are no minimal T-invariants without biological interpretation. Most importantly, there is no known biological behavior without a corresponding—though not necessarily minimal—T-invariant.
- **Validation criterion 3.** All expected special behavioral properties expressed as temporal-logic formulae hold.

One of the benefits of using the qualitative approach is that systems can be modeled and analyzed without any quantitative parameters. Moreover, the qualitative analyses help in identifying suitable initial markings and potential quantitative analysis techniques, as discussed next in Section 3.4.

Now we are ready for a more sophisticated quantitative analysis of our model. Up to this point, we have employed a time-free model. However, the qualitative analysis techniques consider all behavior possible under any timing. Adding timing constraints to our Petri net usually restricts the behavior, so it should not come as a surprise if under the given timing assumptions certain behavior disappears. However, we know that what has not been possible in the qualitative model cannot become possible in the quantitative one—for example, a bounded model cannot become unbounded.

3.4 Quantitative Approaches

In this section we transform our validated qualitative model, given as a place/transition Petri net, into quantitative ones, specified as stochastic or continuous Petri nets, and employ related analysis techniques.

3.4.1 Quantitative Modeling

To transform a qualitative Petri net into quantitative ones, we need to assign to all transitions (reactions) their rate functions, which generally depend on the current state of the reactions' substrates, or—in Petri net terms—the current marking of the transitions' preplaces. Technically, rate functions can be any mathematical functions. However, they often follow some kinetic patterns, with the mass-action kinetics and the Michaelis–Menten kinetics being the most famous ones. Table 3.4 gives for each reaction: the reaction equation, the rate function, and the involved rate constants. The rate functions (employing the stochastic rate constants) are used in the stochastic model as the propensity (hazard) functions, determining the current stochastic firing rates, and in the continuous model as the deterministic rate functions (employing the deterministic rate constants), determining the current deterministic firing rates. The conversion of stochastic and deterministic rate constants into each other is well understood (see, for example, [42]), especially it holds that they are equivalent for first-order reactions. For second-order reactions, we get the stochastic rate constant by multiplying with a scaling factor (level semantics).

Finding the rate constants may turn out to be a difficult and time-consuming process. It usually involves both searches for scientific papers and discussions with the biologists knowledgeable in the pathway under investigation. In our case, we obtain the deterministic rate constants from [9].

In the following discussion, we illustrate the strength of quantitative approaches by a few analysis examples only. We start with the stochastic approach to exclude eccentric system behavior caused by stochastic noise before considering the averaged behavior in the deterministic continuous approach.

Number	Reaction Equation			Rate Function v_i	Rate Constant[a] Stochastic	Deterministic
r1	s1 + s2	→	s3	$c_1 \cdot s1 \cdot s2$	$c'_1 = c_1 \cdot f_s$	$c_1 = 0.53$
r2	s3	→	s1 + s2	$c_2 \cdot s3$	$c'_2 = c_2$	$c_2 = 0.0072$
r3	s3 + s9	→	s4	$c_3 \cdot s3 \cdot s9$	$c'_3 = c_3 \cdot f_s$	$c_3 = 0.625$
r4	s4	→	s3 + s9	$c_4 \cdot s4$	$c'_4 = c_4$	$c_4 = 0.00245$
r5	s4	→	s1 + s5 + s6	$c_5 \cdot s4$	$c'_5 = c_5$	$c_5 = 0.0315$
r6	s5 + s7	→	s8	$c_6 \cdot s5 \cdot s7$	$c'_6 = c_6 \cdot f_s$	$c_6 = 0.8$[a]
r7	s8	→	s5 + s7	$c_7 \cdot s8$	$c'_7 = c_7$	$c_7 = 0.0075$
r8	s8	→	s7 + s9	$c_8 \cdot s8$	$c'_8 = c_8$	$c_8 = 0.071$
r9	s6 + s10	→	s11	$c_9 \cdot s6 \cdot s10$	$c'_9 = c_9 \cdot f_s$	$c_9 = 0.92$
r10	s11	→	s6 + s10	$c_{10} \cdot s11$	$c'_{10} = c_{10}$	$c_{10} = 0.00122$
r11	s11	→	s2 + s10	$c_{11} \cdot s11$	$c'_{11} = c_{11}$	$c_{11} = 0.87$

[a] The stochastic and deterministic rate constants are equivalent for first-order reactions. f_s is a scaling factor to map the given *mass* in the continuous concentration onto a finite number of levels (i.e., tokens), with N being the highest level number, i.e., $f_s = mass/N$.

Table 3.4 The Reaction Equation, Rate Function, and Rate Constants for Each Reaction (Transition). For better readability we use the abbreviations s1...s11 for the involved species. All reactions employ mass-action kinetics. Note that the backward reactions constants are by two orders of magnitude smaller than the constants for the forward reactions. Compare this table with Figure 3.2.

3.4.2 Stochastic Analysis

Like a qualitative Petri net, a stochastic Petri net maintains a discrete number of tokens on its places. But contrary to the time-free case, an exponentially distributed firing rate (waiting time) is associated with each transition r, specified by a parameter λ. Generally, λ is state dependent and defined by a propensity (hazard) function. The firing itself does not consume time, and it again follows the standard firing rule of qualitative Petri nets. Thus all waiting times can theoretically still occur, but the likelihood depends on the probability distribution. Consequently, the system behavior is described by the same discrete state space, and all the different execution runs of the underlying qualitative Petri net can still take place.

More formally, the semantics of a stochastic Petri net (with exponentially distributed waiting times for all transitions) is described by a continuous-time Markov chain (CTMC). The CTMC of a stochastic Petri net without parallel transitions is isomorphic to the reachability graph of the underlying qualitative Petri net, while the arcs between the states are now labeled by the transition rates. This allows for the use of the same powerful analysis techniques for stochastic Petri nets as already applied for qualitative Petri nets.

Table 3.4 provides the details that permit reading the net in Figure 3.2 as a stochastic Petri net, specifying at the same time reaction rate equations (RREs).

1. *Equivalence check by transient analysis.* Due to the isomorphy of reachability graph and CTMC, all qualitative analysis results obtained in the former section are still valid. The influence of time does not restrict the possible system behavior. Specifically, it holds that the CTMC in our case study is reversible, which ensures ergodicity; thus we could start the system in any of the reachable states, always resulting in the same CTMC with the same steady-state probability distribution.

Additionally, probabilistic analyses of the transient and steady-state behavior are now available. Generally, this can be done in an analytic fashion as well as in a simulative manner. The boundedness of the underlying qualitative model allows the application of standard Markov analysis techniques that are based on the state transition matrix. However, stochastic simulation algorithms (SSA), such as Gillespie's exact simulation algorithm [16], are computationally less expensive and work for huge or even infinite state spaces.

Applying SSA produces data describing the dynamic evolution of the biological system over time. Figure 3.6 shows the behavior of a subset of the species in the model (phosphorylated proteins or complexes of them) over time, for various numbers of levels (tokens). The noise decreases as the number of levels increases; thus the behavior of the stochastic Petri net approaches that of the deterministic net. Moreover, it is obvious that the system reaches a steady state in all shown cases, determining the value at which the response saturates.

We proceed with transient analysis to prove the sufficient equivalence between the stochastic model in the level semantics and the corresponding continuous model, justifying the interpretation of the properties gained by the stochastic model also in terms of the continuous one. Probabilistic model checkers such as PRISM [29] and IDD-CSL [36] permit the analysis of the transient behavior of the stochastic model. For example, the concentration of MEK-PP at time t is given by

$$C_{\mathrm{MEK-PP}}(t) = \tfrac{M}{N} \cdot \underbrace{\sum_{i=1}^{N} (i \cdot P(L_{\mathrm{MEK-PP}}(t) = i))}_{\text{Expected value of } \mathrm{L}_{\mathrm{MEK-PP}}(t)}$$

where N is the highest level number and M is the maximum concentration, 2.5 μMol. The random variable $L_{\mathrm{MEK-PP}}(t)$ stands for the level of

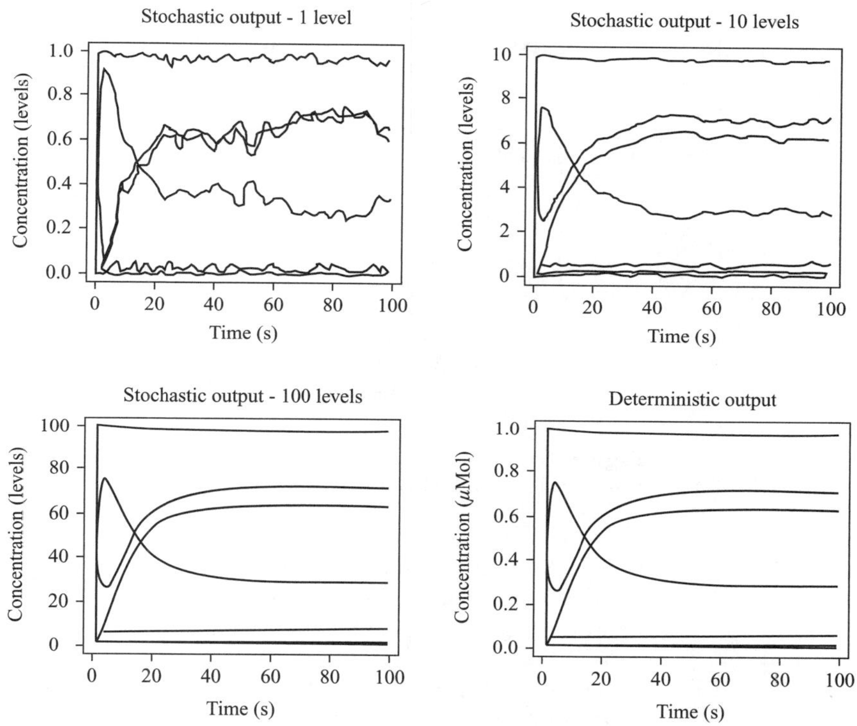

Figure 3.6 Diagrams displaying the output of deterministic (ODE) and stochastic (SSA) simulation for a subset of the species in the system. This shows that the averaged stochastic behavior (over 100 runs) tends toward the deterministic behavior as the number of levels increases. The species shown are the phosphorylated species.

MEK-PP at time t. Figure 3.7 shows the output of MEK-PP and Raf-1Star_RKIP in the time interval $[0 \ldots 100]$ according to the continuous model and the stochastic model, respectively. These results confirm that 7 levels are sufficient to approximate the continuous model, and that there is little discernible difference in output between the number of levels used in the figure.

2. *Analytical stochastic model checking.* In the previous section we employed CTL to express behavioral properties. Given that we now have a stochastic model, we can apply the branching-time Continuous Stochastic Logic (CSL), which replaces the path quantifiers (**E**, **A**) in CTL by the probability operator $\mathbf{P}_{\bowtie p}$, whereby $\bowtie p$ specifies the probability of the given formula. First we introduce in CSL the abbreviations $\mathbf{F}\,\phi$ for $true\,\mathbf{U}\,\phi$

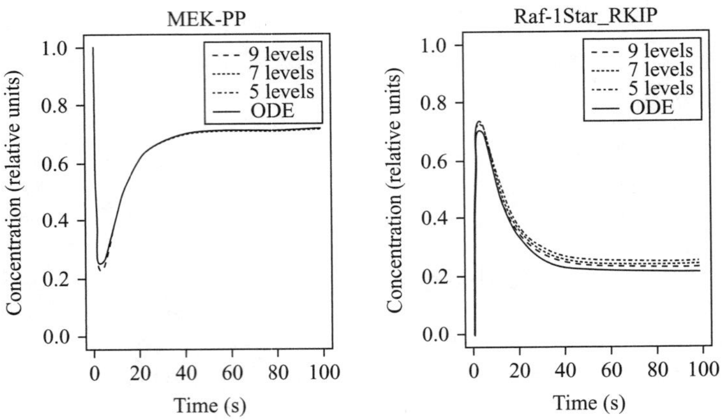

Figure 3.7 Comparison of the concentration traces for 5, 7, and 9 levels in the stochastic model against the continuous behavior.

and $\mathbf{G}\,\phi$ for $!\mathbf{F}\,!\phi$. The CTL formula $\mathbf{EF}\,\phi$ now becomes the CSL formula $\mathbf{P}_{>0}[\,\mathbf{F}\,\phi\,]$, and $\mathbf{AG}\,\phi$ becomes $\mathbf{P}_{\geq 1}[\,\mathbf{G}\,\phi\,]$. CSL also defines a steady-state operator $\mathbf{S}_{\bowtie p}$ that specifies the probability of the given formula in the long run.

The probabilistic model checkers PRISM and IDD-CSL uses CSL logic. Both tools take an analytical approach, where the probabilities in the probability operator are exact and the result of CTMC analysis. IDD-CSL reads a Petri net standard format. To use the PRISM model checker, we have to encode our running example in its modeling language. This translation requires knowledge of the boundedness degree of all species involved, which we acquire by the structural analysis technique of P-invariants. We consider here the level semantics, where we divide the concentration range (0...2.5 μMol) into $N + 1$ discrete levels. Thus with M, the maximum concentration being 2.5 μMol, we define levels as such: $0, (0 * M/N, 1 * M/N], (1 * M/N, 2 * M/N], ...((N-1) * M/N, N * M/N]$. The discrete tokens in the stochastic Petri net now relate to the levels. We use the result from transient analysis that 7 levels are sufficient to describe the behavior of the system and continue with a 7-level model.

We can now check our stochastic Petri net with CSL properties equivalent to the qualitative CTL properties described in the previous section, whereby places are interpreted as integer variables.

- **Property S1a:** There are reachable states where ERK is phosphorylated and RKIP is not phosphorylated.

$$\mathbf{P}_{>0}\,[\,\mathbf{F}\,(\text{ERK-PP} + \text{Raf-1Star\_RKIP\_ERK-PP} \geq 1 \wedge \text{RKIP} + \text{Raf-1Star\_RKIP\_ERK-PP} + \text{Raf-1Star\_RKIP} = 7\,)]$$

- **Property S2a:** The phosphorylation of ERK (to ERK-PP) is independent of the phosphorylated state of RKIP.

$$\mathbf{P}_{=?}\,[\mathbf{G}\,((\text{ERK} = 7 \wedge \text{RKIP-P} = 0 \wedge \text{RKIP-P\_RP} = 0) \rightarrow \mathbf{P}_{>0}\,[(\text{RKIP-P} = 0 \wedge \text{RKIP-P\_RP} = 0)\,\mathbf{U}(\text{ERK-PP} \geq 1)])]$$

  We take this opportunity to use the $\mathbf{P}_{=?}$ operator, which returns (rather than compares) the probability of the property. As this property always holds true, the returned probability is 1.

- **Property S3a:** A cyclic behavior w.r.t. the presence/absence of RKIP is possible forever.

$$\mathbf{S}_{>0}\,[(\text{RKIP}{=}\,0)]$$
$$\mathbf{S}_{>0}\,[(\text{RKIP}{\geq}\,1)]$$

  This property is expressed using the steady-state operator, which states that in the long run there are non-zero probabilities of RKIP being present or absent; hence it can cycle.

Analytical model checking constructs and analyzes the entire state space—an effort that can become infeasible due to the state explosion problem. Thus we now move to a simulative approach.

3. *Simulative stochastic model checking.* A simulative approach handles large state spaces through approximating results by analyzing only a subset of the state space—a finite set of finite paths through the CTMC produced by a simulation algorithm.

The type of logic now suitable for describing properties changes from branching time (e.g., CSL operating over the CTMC) to linear time. A linear-time logic operates, in turn, over paths through the CTMC, equivalent to operating on simulation outputs. We use a probabilistic linear-time temporal logic (PLTL) that extends standard LTL [31] to a stochastic setting. PLTL is LTL extended with a $\mathbf{P}_{\bowtie p}$ operator and a filter construct, $\{\phi\}$, defining the initial state from which the property is checked. Because PLTL is a linear-time logic, there can be no embedded probability operators.

We use the PLTL Monte Carlo model checker, MC2 [26], with Gillespie's algorithm as featured in Snoopy. The input to the model checker was 100 simulation outputs from Gillespie's algorithm, with a simulation time of 10,000 s. We accept the result that 7 tokens are sufficient and use 7 molecules in the simulation. We now check the following PLTL properties:

- **Property S1s:** There are reachable states where ERK is phosphorylated and RKIP is not phosphorylated.

$$\mathbf{P}_{>0}\,[\mathbf{F}\,(\text{ERK-PP} + \text{Raf-1Star\_RKIP\_ERK-PP} \geq 1 \wedge \text{RKIP} + \text{Raf-1Star\_RKIP\_ERK-PP} + \text{Raf-1Star\_RKIP} = 7)]$$

- **Property S2s:** The phosphorylation of ERK (to ERK-PP) is independent of the phosphorylated state of RKIP.

  $\mathbf{P}_{\geq 1}$ [(RKIP-P + RKIP-P_RP = 0) **U** (ERK-PP $\geq$ 1)]

  Because PLTL cannot have embedded probability operators, we are restricted here to a much weaker property. This property says from the initial state of the system, that ERK can phosphorylate without RKIP being phosphorylated. In fact, from the initial state this is always true.
- **Property S3s:** A cyclic behavior w.r.t. the presence/absence of RKIP is possible forever.

  $\mathbf{P}_{>0}$ [**F** (RKIP $\geq$ 1){time $\geq$ 5000 $\wedge$ RKIP = 0}]

  $\mathbf{P}_{>0}$ [**F** (RKIP = 0){time $\geq$ 5000 $\wedge$ RKIP $\geq$ 1}]

  PLTL cannot directly perform steady-state analysis. Steady-state behavior can be approximated through long simulations with the property checked toward the end of the simulation. The filter, $\{\phi\}$, defines the initial state that the property is checked from. Late in the simulation where RKIP is absent, we check whether RKIP can become present, and vice versa. This property also holds when a time greater than 5000 s is used.

Simulative stochastic model checking has its advantage when the range of behaviors that the model checker operates over becomes computationally infeasible. A simulative approach will allow model checking results in a feasible time through approximation. In contrast, our stochastic Petri net with 7 tokens is feasible in both approaches, and we would suggest that analytical approaches should be used where possible.

We continue with the computationally less expensive continuous approach considering the averaged behavior.

3.4.3 Continuous Analysis

In a continuous Petri net, the marking of a place is no longer an integer, but a non-negative real number, which can be read as the concentration of the species modeled by the place. Also, transitions fire continuously according to the deterministic rate functions. Assigning these rate functions to all transitions (see Table 3.4) allows us to read the net in Figure 3.2 as a continuous Petri net.

The semantics of a continuous Petri net is defined by a system of ordinary differential equations (ODEs), whereby one equation describes the continuous change over time on the token value of a given place by the continuous increase of its pre-transitions' flow and the continuous decrease of its post-transitions' flow. In other words, each place p subject to changes gets its own equation. The complete system of non-linear ODEs generated from the continuous Petri net of our running example is described next. Note that the place names now stand for real number variables, and the rate function variables v_i are specified in Table 3.4. To make the model-checking analysis more intuitive, we have mapped the concentrations of 2.5 μMol to 1 μMol such that all concentrations are relative.

$$\frac{ds_1}{dt} = v_2 + v_5 - v_1 \qquad \frac{ds_5}{dt} = v_5 + v_7 - v_6 \qquad \frac{ds_9}{dt} = v_4 + v_8 - v_3$$

$$\frac{ds_2}{dt} = v_2 + v_{11} - v_1 \qquad \frac{ds_6}{dt} = v_5 + v_{10} - v_9 \qquad \frac{ds_{10}}{dt} = v_{10} + v_{11} - v_9$$

$$\frac{ds_3}{dt} = v_1 + v_4 - v_2 - v_3 \qquad \frac{ds_7}{dt} = v_7 + v_8 - v_6 \qquad \frac{ds_{11}}{dt} = v_9 - v_{10} - v_{11}$$

$$\frac{ds_4}{dt} = v_3 - v_4 - v_5 \qquad \frac{ds_8}{dt} = v_6 - v_7 - v_8$$

1. *Steady-state analysis.* Simulating the continuous Petri net (i.e., solving numerically the underlying system of ODEs), we get data as shown in Figure 3.9. Given the usual inexactitudes of the rate constants, biochemists will often interpret the results of ODE-based simulations as indicators of the behavior of the components of the network, rather than being concerned with the exact value of the concentration of a particular species at a particular point in time.

We perform a quantitative analysis of the simulation results to determine whether the 13 "good" initial states (markings) suggested by the qualitative Petri net analysis (see Table 3.3) are, indeed, in some way equivalent (they all result in the same steady state), and that no other possible initial states can be used to give the same results. These are "sensible" initial states from the point of view of biochemistry, in that in all these 13 cases, and in none of the other 2035 states, each protein species is in a high initial concentration in only one of the following states: uncomplexed, complexed, unphosphorylated, or phosphorylated. These conditions relate exactly to the 1-P-invariant interpretation given in our initial marking construction procedure in Section 3.3.2.

For the purposes of our computations, we map any non-zero concentrations to 1. We then compute the steady state of the set of species for each possible initial state—for example, using the MATLAB ODE solver ode45, which is based on an explicit Runge-Kutta formula, the Dormand-Prince pair [12], with 100 time steps. We find that all of the 13 "good" initial states result in the same final state, within the bounds of computational error of the ODE solver, while none of the other 2035 "bad" initial states results in a steady state near that of the set of "good" states (compare Figure 3.8).

We reproduce two simulations of the model: State 1 (Figure 3.9, left), which corresponds to the initial marking suggested by our approach

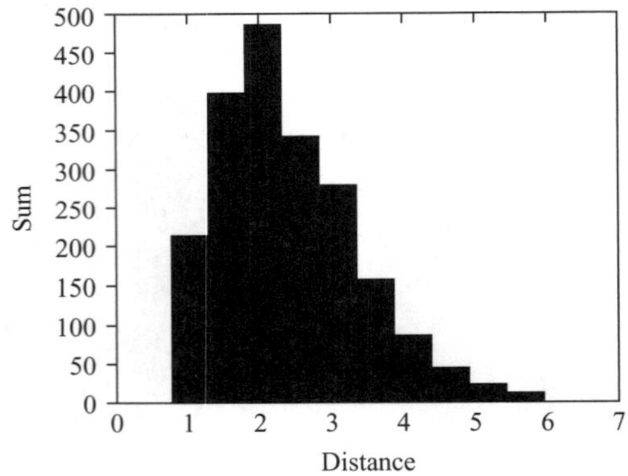

Figure 3.8 Distribution of the Euclidean distances between the steady-state vector of the "good" initial states and each of the steady-state vectors for the "bad" initial states. The distances range from 0.7736 to 6.0889.

described in Section 3.3.2 with ERK-PP low and ERK high, and State 5 (Figure 3.9, right), which corresponds to the initial marking suggested by Cho et al. [9], where the initial concentration of ERK-PP is high and ERK is low. State 1 has been confirmed by an expert signal transduction researcher as the most sensible starting state [23]. The equivalence of the steady states, compared with the difference in some intermediate states, is clearly illustrated in these figures. For example, the concentration of Raf-1Star_RKIP behaves overall in a similar manner in both State 1 and State 5, peaking before 10 seconds—although the peak is greater when ERK is not phosphorylated at the start of the experiment. Figure 3.10 reproduces the computed behavior of ERK-PP for all 13 good initial states. Despite differences in the concentrations at early time points, the steady-state concentration is the same in all 13 states.

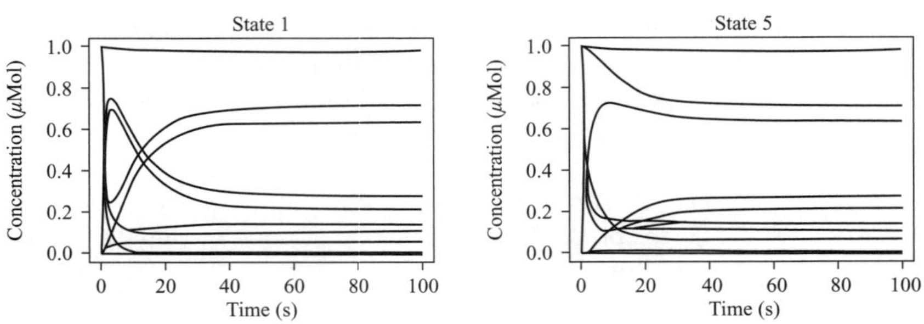

Figure 3.9 Dynamic behavior for State 1 (m1) and State 5 (m5).

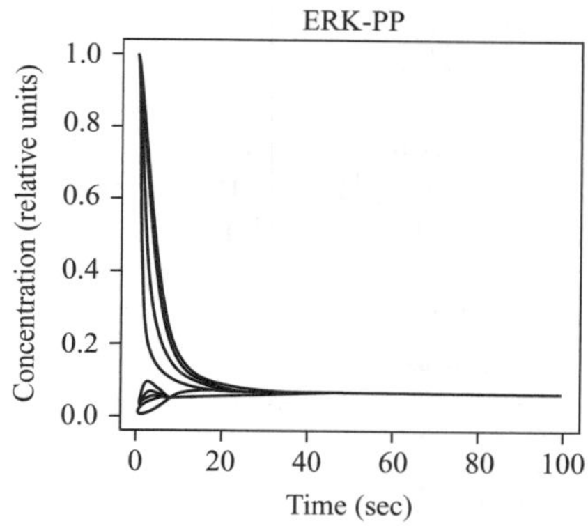

Figure 3.10 Dynamic behavior of ERK-PP for all 13 "good" states

2. *Continuous model checking of the transient behavior.* The continuous model replaces the discrete values of species with continuous values, and hence is not able to describe the behavior of species at the level of individual molecules, but only the overall deterministic behavior via concentrations. The concentration of a particular species will have the same value at each point of time for repeated experiments. The state space of such models is continuous and linear. For all these reasons, a linear-time logic is a natural choice to analyze such kind of behavior.

We employ the PLTL-based MC2 model checker again as per the stochastic analysis. MC2 operates on a deterministic model in the same manner as on a stochastic model—analyzing simulation output. However, in the deterministic model there is only one behavior (the average behavior), so the resulting probability is 1 or 0. To provide simulation output to MC2, we use the BioNessie simulator [3] with a simulation time of $10,000\,\mathrm{s}$.

The PLTL properties from the stochastic analysis are rewritten to concentration values, comparing to real numbers called *noise* and *significant*. The place names are now interpreted as real number variables.

- **Property C1:** There are reachable states where ERK is phosphorylated and RKIP is not phosphorylated.

  $$\mathbf{P}_{\geq 1}\,[\mathbf{F}\,(\text{ERK-PP} + \text{Raf-1Star\_RKIP\_ERK-PP} > \mathit{noise} \wedge \text{RKIP} + \text{Raf-1Star\_RKIP\_ERK-PP} + \text{Raf-1Star\_RKIP} \geq \mathit{significant})]$$

- **Property C2:** The phosphorylation of ERK (to ERK-PP) is independent of the phosphorylated state of RKIP.

  $$\mathbf{P}_{\geq 1}\,[(\text{RKIP-P} + \text{RKIP-P\_RP} \leq \mathit{noise})\ \mathbf{U}\ (\text{ERK-PP} + \text{Raf-1Star\_RKIP\_ERK-PP} > \mathit{noise})]$$

  Due to the continuous change of token values and the large kinetic rate of ERK-PP's post-transition relative to its pre-transitions, ERK-PP's concentration quickly becomes consumed in the complex

Raf-1Star_RKIP_ERK-PP. Therefore, this complex has been added to the property.

- **Property C3:** A cyclic behavior w.r.t. the presence/absence of RKIP is possible forever.

  $\mathbf{P}_{\leq 0}$ [**F** (RKIP $>$ *noise*){ time $\geq$ 5000 $\wedge$ RKIP $\leq$ *noise* }]

  $\mathbf{P}_{\leq 0}$ [**F** (RKIP $\leq$ *noise*){ time $\geq$ 5000 $\wedge$ RKIP $>$ *noise* }]

  As this is the average behavior of the system, the cyclic behavior of RKIP is averaged and is no longer present. Hence Property C3 says that it is not true ($\mathbf{P}_{\leq 0}$) that RKIP can cycle between noise and not noise.

These properties hold for the intuitive values of 0.1 μMol and 0.9 μMol for *noise* and *significant*, respectively.

3.4.4 Summary

In summary, quantitative Petri nets can provide interesting analysis in two complementary worlds—stochastic and continuous. The key points that follow have been discussed in this section:

1. The qualitative Petri net can easily be converted to quantitative Petri nets by the addition of rate equations.
2. All of the 13 states identified by the reachability graph of the qualitative Petri net result in the same set of steady-state values for the 11 species in the pathway.
3. None of the remaining 2035 possible initial states of the qualitative Petri net results in a final steady state close to that generated by the 13 markings in the reachability graph.
4. The transient behavior—the crucial point of interest in signal transduction networks—of the quantitative models is sensible for the 13 states identified by our method.
5. There are significant differences between the three quantitative model-checking approaches. Analytical model checking provides exact analysis but quickly becomes infeasible for a larger number of levels. Simulative model checking provides analysis beyond that which analytical model checking can achieve through approximation, but there can be restrictions on the properties that can be checked. Continuous model checking analyzes the average behavior of the system only.
6. Model checking the stochastic and continuous Petri nets complement each other; it is possible that there are behaviors in the stochastic Petri net that are lost in the continuous net. One should model check the stochastic Petri net before considering the overall behavior of the system in the continuous Petri net.

3.5 Tools

The running example in its three interpretations has been done using Snoopy—a tool to design and animate or simulate hierarchical graphs, among them the qualitative, stochastic, and continuous Petri nets used in this chapter. Snoopy provides export to various analysis tools as well as SBML import and export [18, 38].

The qualitative analyses have been made with the Petri net analysis tool Charlie [8] and the Model Checking Kit [35], supported by IDD-CTL, a CTL model checker utilizing interval decision diagrams for concise state-space representations [20, 41].

The quantitative analyses have been done using Snoopy's built-in simulation algorithms for stochastic and continuous Petri nets, and by BioNessie [3], an SBML-based simulation and analysis tool for biochemical networks. Additionally, MATLAB [37] was used to produce the steady-state analysis of all initial states in the continuous model. We employed PRISM [29] and IDD-CSL [36] for analytical stochastic model checking, and MC2 [26] for simulative stochastic and continuous model checking. We use the Gillespie simulator in Snoopy as input to stochastic MC2 and the BioNessie simulator as input to continuous MC2.

More Petri net tools and related materials can be found on the Petri Nets World's website: `http://www.informatik.uni-hamburg.de/TGI/PetriNets/`.

3.6 Further Reading

General introductions into Petri net theory can be found in [27, 34, 39]; for stochastic Petri nets, see specifically [1, 2]; for continuous and hybrid Petri nets, see [11].

Biochemically interpreted stochastic Petri nets have been introduced in [14]. An extension of this net class is used in [19] for the model-based design of wetlab experiments employing simulative stochastic model checking. Biochemically interpreted continuous Petri nets—as used in this chapter—have been introduced in [13], which are deployed in [4] for a structured approach for the engineering of biochemical network models.

Recent surveys on applying Petri nets for biochemical networks are [7, 25], offering a rich choice of further reading pointers, among them numerous case studies applying Petri nets to biochemical networks. Besides the net classes introduced in this chapter, colored Petri nets, duration and interval time Petri nets, and hybrid Petri nets in various extensions have been employed.

P- and T-invariants have been established concepts in Petri net theory since the very beginning [24]. Corresponding notions can be found in systems biology, called chemical moieties or conservation relations, and elementary modes or extreme pathways which are elaborated in the setting of biochemical networks in [28]. For biochemical systems without reversible reactions, the notions of T-invariants, elementary modes, and extreme pathways coincide.

A good starting point for qualitative model checking is [10]. For numerical solution of Markov chains, consult [40]. An overview on stochastic issues for systems biology is given in [42].

The systematic qualitative analysis of a metabolic network is demonstrated in [22], following basically the same outline as used in Section 3.3.2. Further examples illustrating how to combine qualitative and quantitative analysis techniques are elaborated in [15, 17]. The first paper also provides the formal definitions of the technical Petri net terms used in this chapter as well as further references to related papers.

SUMMARY

In this chapter we introduced qualitative, stochastic, and continuous Petri nets as well as related analysis techniques in a deliberately informal manner. Our motivation has been to make the concepts accessible for both modelers and biologists who are not familiar with this methodology. For the sake of biological coherence, we have used a signaling pathway—namely, the influence of Raf Kinase Inhibitor Protein (RKIP) on the Extracellular signal Regulated Kinase (ERK) signaling pathway [9], as a running example. However, the techniques described in this chapter can be equally applied to other types of biochemical networks in systems biology or synthetic biology—for example, metabolic or gene transcription networks, or even combinations of them.

The nature of biochemical reaction systems—bipartite (in terms of species and interactions), concurrent, and stochastic—means that Petri nets are a natural choice for a formal modeling technique to describe their structure and behavior. The qualitative aspects of these systems, such as the biochemical entities, and their relationships up to stoichiometry can be modeled with qualitative Petri nets, whereas their quantitative behavior can be described by stochastic Petri nets at the individual (e.g., molecular) level and continuous Petri nets on the larger scale. Petri nets are a well-established technique for modeling computational systems that are highly suitable for biochemical reaction systems. A large range of robust Petri net modeling tools are available; for our work, we have chosen Snoopy [18, 38], which is freely obtainable and is available for the major computational platforms. As well as being able to describe static information about biochemical reaction systems, Petri nets are able to simulate their behavior using, for example, the token game for qualitative nets, stochastic simulation algorithms for stochastic nets, and ordinary differential equation solvers for continuous nets.

In addition to description and simulation, very powerful techniques are available to analyze the behavior of such systems. Qualitative Petri nets can be analyzed based on their structural or behavioral properties using classical techniques. Model checking can be used to prove special properties that reflect the intended functionality of the network. This can be done for all three types of nets—qualitative, stochastic, and continuous—using the appropriate temporal logics, and we have chosen to use model checking of properties of our running example as the unifying thread throughout this paper.

In summary, we advocate a three-step technology for the modeling and analysis of biochemical networks in a systematic manner: (1) qualitative—that is, (time-free) qualitative modeling and analysis, especially for the beneficial effect of confidence-increasing model validation; (2) stochastic modeling and analysis to exclude eccentric behavior; and (3) continuous modeling and analysis, especially with the hope of the reliable prediction of behavior in the averaged case. For all steps, we favor the deployment of both qualitative and quantitative Petri nets, sharing the same net structures for a given case. The quantitative models are derived from the qualitative one only by the addition of the quantitative parameters. Hence these models are likely to share some behavioral properties.

The running example in its three versions and all related analysis protocols are available at the following website: `www-dssz.informatik.tu-cottbus.de/examples/erk`.

EXERCISES

You should be able to solve the first three tasks without tool support, while the remaining tasks suggest the employment of adequate tools.

1. The basic Petri net properties—boundness, liveness, and reversibility—are orthogonal; that is, they are independent from each other. Prove this statement by providing a net for each possible combination (obviously, there are eight of them). Argue for every net why the properties hold as you claim.

If you found this task too easy, repeat this exercise with the additional restriction that the nets should be (a) ordinary and (b) strongly connected.

2. A well-defined matrix operation is transposition that exchanges rows and columns. If we apply the matrix transposition to the incidence matrix of a Petri net, what happens with the Petri net and its invariants?

3. While CTL model checking fits particularly well with making decisions about special properties, it can also be used to decide general properties. Express the following properties in CTL:

(a) Check the liveness of reactions r1 and r9. What had to be done to decide the liveness of the net?

(b) Check the reversibility of the net.

(c) Check the token conservation of the P-invariants.

(d) The CTL Property 2 (see Section 3.3.2) does not hold in its stronger version, where EU is replaced by AU, not even for the reduced network neglecting the backward directions of reversible reactions. Why?

4. Extend the running example of this chapter by the following reactions [5]:

r12, r13:	MEK	+	Raf-1*	$\leftrightarrow$	MEK_RAF-1*	
r14:	MEK_RAF-1*			$\rightarrow$	MEK-PP	+ Raf-1*
r15:	MEK-PP			$\rightarrow$	MEK.	

Adapt the given Petri net model and apply all analysis techniques presented in this chapter in a stepwise manner, following exactly the same outline. Use the following rate constants: $k_{12} = 0.05$, $k_{13} = 0.03$, $k_{14} = 0.06$, $k_{15} = 0.02$.

Discuss also the version where the backward reactions of all reversible reactions are neglected: Interesting behavior, isn't it?

5. Use the techniques you have learned in this chapter to

(a) Derive qualitative and quantitative Petri net models for the two-stage signal transduction cascade given in Figure P.5.

(b) Construct a suitable initial marking.

(c) Animate and simulate their behaviors.

(d) Analyze their properties and behaviors.

Use the following rate constants for the quantitative models:
$k_1 = kr_1 = kk_1 = kkr_1 = 100$
$k_2 = kr_2 = kk_2 = kkr_2 = 4$
$k_3 = kr_3 = kk_3 = kkr_3 = 1$

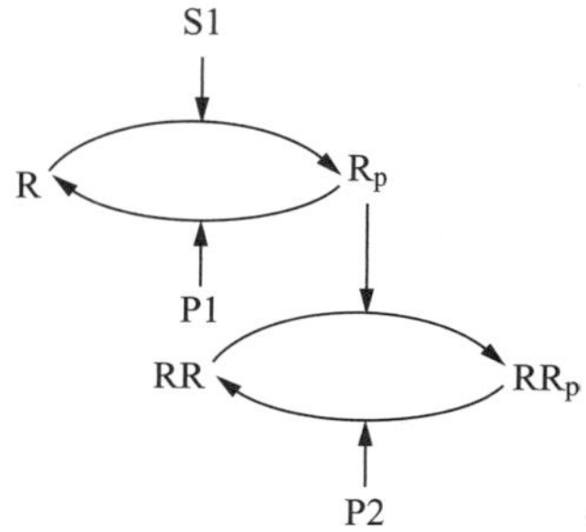

$$R + S_1 \underset{k_2}{\overset{k_1}{\rightleftarrows}} R|S_1 \overset{k_3}{\longrightarrow} R_p + S_1$$

$$R_p + P_1 \underset{kr_2}{\overset{kr_1}{\rightleftarrows}} R_p|P_1 \overset{kr_3}{\longrightarrow} R + P_1$$

$$RR + R_p \underset{kk_2}{\overset{kk_1}{\rightleftarrows}} RR|R_p \overset{kk_3}{\longrightarrow} RR_p + R_p$$

$$RR_p + P_2 \underset{kkr_2}{\overset{kkr_1}{\rightleftarrows}} RR_p|P_2 \overset{kkr_3}{\longrightarrow} RR + P_2$$

Figure P.5 The two-stage signal transduction cascade

References

[1] M. Ajmone Marsan, G. Balbo, G. Conte, S. Donatelli, and G. Franceschinis. *Modelling with Generalized Stochastic Petri Nets.* 2nd ed. Wiley Series in Parallel Computing, John Wiley and Sons, 1995.

[2] F. Bause and P. S. Kritzinger. *Stochastic Petri Nets.* Vieweg, 2002.

[3] BioNessie website. A biochemical pathway simulation and analysis tool. University of Glasgow, `http://www.bionessie.org`, 2008.

[4] R. Breitling, D. Gilbert, M. Heiner, and R. Orton. A structured approach for the engineering of biochemical network models, illustrated for signalling pathways. *Briefings in Bioinformatics,* (9):404–421, 2008.

[5] M. Calder, S. Gilmore, and J. Hillston. Automatically deriving ODEs from process algebra models of signalling pathways. In *Proceedings of the CSMB 2005*, pages 204–215. LFCS, University of Edinburgh, 2005.

[6] N. Chabrier-Rivier, M. Chiaverini, F. Fages V. Danos, and V. Schächter. Modeling and querying biomolecular interaction networks. *Theoretical Computer Science*, 325(1):25–44, 2004.

[7] C. Chaouiya. Petri net modeling of biological networks. *Briefings in Bioinformatics*, 8(4):210–219, 2007.

[8] Charlie website. A tool for the analysis of place/transition nets. BTU Cottbus, http://www-dssz.informatik.tu-cottbus.de/software/charlie/charlie.html, 2008.

[9] K.-H. Cho, S.-Y. Shin, H.-W. Kim, O. Wolkenhauer, B. McFerran, and W. Kolch. Mathematical modeling of the influence of RKIP on the ERK signaling pathway, volume 2602 of *Lecture Notes in Computer Science,* pages 127–141, Springer, 2003.

[10] E. M. Clarke, O. Grumberg, and D. A. Peled. *Model checking.* MIT Press, 2001.

[11] R. David and H. Alla. *Discrete, Continuous, and Hybrid Petri Nets.* Springer, 2005.

[12] J. R. Dormand and P. J. Prince. A family of embedded Runge-Kutta formulae. *Journal of Computational and Applied Mathematics*, 6:1–22, 1980.

[13] D. Gilbert and M. Heiner. From Petri nets to differential equations—an integrative approach for biochemical network analysis. In *Proceedings ICATPN 2006*, volume 4024 of *Lecture Notes in Computer Science*, pages 181–200. Springer, 2006.

[14] D. Gilbert, M. Heiner, and S. Lehrack. A unifying framework for modelling and analysing biochemical pathways using Petri nets. In *Proceedings CMSB 2007*, volume 4695 of *Lecture Notes in Computer Science/LNBI*, pages 200–216. Springer, 2007.

[15] D. Gilbert, M. Heiner, S. Rosser, R. Fulton, Xu Gu, and M. Trybiło. A case study in model-driven synthetic biology. *2nd IFIP Conference on Biologically Inspired Collaborative Computing (BICC), IFIP WCC 2008, Milano, pages 163–175*, 2008.

[16] D. T. Gillespie. Exact stochastic simulation of coupled chemical reactions. *Journal of Physical Chemistry*, 81(25):2340–2361, 1977.

[17] M. Heiner, D. Gilbert, and R. Donaldson. Petri nets in systems and synthetic biology. In *Schools on Formal Methods (SFM)*, volume 5016 of *Lecture Notes in Computer Science*, pages 215–264. Springer, 2008.

[18] M. Heiner, R. Richter, and M. Schwarick. Snoopy: A tool to design and animate/simulate graph-based formalisms. In *Proceedings of PNTAP 2008*. ACM ISBN 978-963-9799-20-2, 2008.

[19] M. Heiner, S. Lehrack, D. Gilbert, and W. Marwan. Extended stochastic Petri nets for model-based design of wetlab experiments, volume 5750 of *Lecture Notes in Computer Science/LNBI*, pages 138–163. Springer, 2009.

[20] M. Heiner, M. Schwarick, and A. Tovchigrechko. DSSZ-MC—A Tool for symbolic analysis of extended Petri nets. In *Proceedings Petri Nets 2009*, volume 5606 of *Lecture Notes in Computer Science*, pages 323–332. Springer, 2009.

[21] R. Hofestädt. A Petri net application of metabolic processes. *Journal of System Analysis, Modeling and Simulation*, 16:113–122, 1994.

[22] I. Koch and M. Heiner. Petri nets. In B. H. Junker and F. Schreiber, editors, *Biological Network Analysis*, pages 139–179. Wiley Book Series on Bioinformatics, John Wiley and Sons, 2008.

[23] Walter Kolch, personal communication, 2005.

[24] K. Lautenbach. *Exact Liveness Conditions of a Petri Net Class* [in German]. Technical report, GMD Report 82, Bonn, 1973.

[25] H. Matsuno, C. Li, and S. Miyano. Petri net based descriptions for systematic understanding of biological pathways. *IEICE—Transactions on Fundamentals of Electronics, Communications and Computer Sciences*, E89-A(11):3166–3174, 2006.

[26] MC2 website. MC2: PLTL model checker. University of Glasgow, `http://www.brc.dcs.gla.ac.uk/software/mc2/`, 2008.

[27] T. Murata. Petri nets: Properties, analysis and applications. *Proceedings of the IEEE 77*, 4:541–580, 1989.

[28] B. O. Palsson. *Systems Biology: Properties of Reconstructed Networks.* Cambridge University Press, 2006.

[29] D. Parker, G. Norman, and M. Kwiatkowska. *PRISM 3.0.beta1 Users' Guide*, 2006.

[30] C. A. Petri. *Communication with Automata* [in German]. Schriften des Instituts für Instrumentelle Mathematik, Bonn, 1962.

[31] A. Pnueli. The temporal semantics of concurrent programs. *Theoretical Computer Science*, 13:45–60, 1981.

[32] V. N. Reddy. *Modeling Biological Pathways: A Discrete Event Systems Approach.* Master's Thesis, University of Maryland, 1994.

[33] V. N. Reddy, M. L. Mavrovouniotis, and M. L. Liebman. Petri net representations in metabolic pathways. In *Proceedings of the International Conference on Intelligent Systems for Molecular Biology*, 1993.

[34] W. Reisig. *Petri Nets: An Introduction.* Springer, 1982.

[35] C. Schröter, S. Schwoon, and J. Esparza. The model checking kit. In *Proceedings of ICATPN*, volume 2679 of *Lecture Notes in Computer Science*, pages 463–472. Springer, 2003.

[36] M. Schwarick and M. Heiner. CSL model checking of biochemical networks with interval decision diagrams. In *Proceedings 7th CMSB 2009*, volume 5688 of *Lecture Notes in Computer Science/LNBI*, pages 296–312. Springer, 2009.

[37] L. F. Shampine and M. W. Reichelt. The MATLAB ODE suite. *SIAM Journal on Scientific Computing*, 18:1–22, 1997.

[38] Snoopy website. A tool to design and animate/simulate graphs. BTU Cottbus, http://www-dssz.informatik.tu-cottbus.de/software/snoopy.html, 2008.

[39] P. H. Starke. *Analysis of Petri Net Models* [in German]. B. G. Teubner, Stuttgart, 1990.

[40] W. J. Stewart. *Introduction to the Numerical Solution of Markov Chains.* Princeton University Press, 1994.

[41] A. Tovchigrechko. *Model Checking Using Interval Decision Diagrams.* PhD Thesis, BTU Cottbus, Department of Computer Science, 2008.

[42] D. J. Wilkinson. *Stochastic Modelling for System Biology.* CRC Press, 2006.

Chapter

4

On Statecharts for Biology

Jasmin Fisher and David Harel

Jasmin Fisher, Microsoft Research Cambridge, Cambridge, UK
David Harel, Department of Computer Science & Applied Mathematics, Weizmann Institute of Science, Rehovot, Israel

Abstract We discuss recent work carried out on using the Statecharts language for modeling biological systems. The motivation comes from (1) the similarity between the behavioral complexity involved in understanding biology and that encountered when developing reactive systems, and (2) the advantages offered by intuitive visual languages for capturing behavior.

4.1 Biology as Reactivity

One of the central issues in software and system engineering over the last couple of decades has been the development of languages, methods, and tools for the reliable construction of reactive systems—those whose complexity stems not necessarily from complicated computation but from complicated reactivity over time [19]. Such systems are usually highly concurrent and time intensive, and can exhibit hybrid behavior that is predominantly discrete in nature but has continuous and stochastic aspects too. The structure of these systems consists of many interacting, often distributed, components. Very often the structure itself is dynamic, with its components being repeatedly created and destroyed during the system's life span. The heart of the problem is the need for good approaches to modeling and analyzing the dynamic behavior of such systems.

In the last decade or so, we and others have carried out work on viewing biological systems as reactive systems, and on modeling and analyzing a variety of sample pieces of biology. This approach is based on three premises [16]: (1) that satisfactory frameworks exist in software and systems engineering for reactive system modeling and design, (2) that biological research is ready for an extremely significant transition from analysis (reducing experimental observations to elementary building blocks) to synthesis (integrating the parts into a comprehensive whole)—a transition that requires mathematics and computation, and (3) that the true complexity of biological systems—specifically multicellular living organisms—stems from their reactivity.

As far as the third of these suppositions goes, biological systems exhibit the characteristics of reactive systems remarkably, and on many levels—from the molecular level, via the cellular system, all the way up to organs, full organisms, and even entire populations. It doesn't take much to observe within such systems the heavy concurrency, the event-driven discrete nature of the behavior, the chain reactions and cause–effect phenomena, the time-dependent patterns, and so on. From these three premises, there follows a thesis, prompting much of our own work in this area: that biological systems can be modeled beneficially as reactive systems, using languages and tools developed in computer science for the construction of human-made systems. One of the main languages we have used to do the modeling is that of Statecharts [14, 17], and the present chapter is aimed at discussing some of this work and its ramifications.

Our claim is that biological systems, just like many engineered systems, are complex reactive systems interacting with and responding to the environment and to other components of the system [16]. The main efforts seeking to model reactive biological systems with the aid of Statecharts include the work carried out by Kam et al. [20] on T-cell activation, and Efroni et al. [9] on T-cell development in the thymus. Following work on a model of cell fate determination during *C. elegans* vulval development by Kam et al. [21], who used a somewhat different language (Live Sequence Charts), Statecharts were used to model certain advanced aspects of that system by Fisher et al. [12, 13] and Sadot et al. [26]. Swerdlin et al. [29] have modeled the dynamics of the lymph node, and Setty et al. [27] have modeled the morphogenesis of the pancreas.

To help see how reactivity is indeed relevant, consider the following description of the reactivity of biology.

A reactive system, in contrast to a transformational system, does not behave according to a preprogrammed chain of linked instructions. Rather, such a system reacts in parallel to many concurrent inputs. Its behaviors, outputs, and effects are a function not just of the values of its inputs, but also of their variety, of the order in which they arrive, of their timing, of their arrival speeds, and so forth. A biological system expresses a dynamic narrative in which the DNA code is one of the many formative inputs. Structural proteins, enzymes, carbohydrates, lipids, intracellular signals, hormones, and other molecules play key roles in forming and informing the system. The environment of the living system is a most critical source of information [5]. True, DNA serves as a special repository of information because it is replicated and transmitted across generations, but DNA is meaningless without the proteins and other molecules that selectively activate segments of the DNA sequence in variable and alternative ways to create genes. The activation of specific genes emerges from the dynamic state of the cell. One could argue that DNA is just as much a servant of the cell's state as it is the cell's master: There is no hierarchical master plan [6].

Unlike a transformational system, a reactive system does not seek equilibrium, and it has no set point and no state of rest. A reactive system holds itself together as a system just by reacting. It succeeds not by reaching homeostasis; a brain in homeostasis is clinically dead. Rather, a reactive system succeeds by being both robust and resilient. The reactive system responds to simultaneous perturbations and continues to survive, thanks to its reactive dynamics. It is true that organisms feature subsystems that can be described nicely by homeostatic principles; examples include the thermoregulatory system that maintains our internal temperature at 37 °C, our water balance system, and blood glucose regulation. But these systems act like the transformational systems only when viewed as a whole; internally, each of these systems is fashioned by collectives of concurrently reactive, never-resting cells [4].

4.2 Visual Languages for Reactivity

Dynamic reactive behavior is the main issue of concern in this chapter. Note that we will have nothing much to say about how the structural, non-dynamic aspects of a biological system are to be captured.

One of the most widely used ideas that has proven relevant to the process of specifying the dynamic behavior of reactive systems is that of visual formalisms [15]. These languages are both graphically intuitive and mathematically rigorous, and many are supported by powerful tools that support full-model executability and code generation. The models they give rise to are linkable to graphical user interfaces (GUIs) and other structural descriptions of the system under development. This capability facilitates realistic simulation prior to actual implementation. Recently, many of these languages and tools have been based on the object-oriented paradigm and are being strengthened by verification modules, making it possible not only to execute and simulate the system models (test and observe) but also to verify the dynamic properties of those models (prove). They are also being linked to tools for dealing with the system's continuous aspects in a full-hybrid fashion. Most of the tools are state based, encouraging intra-object-style specification, and within these the language of choice for reactive behavior is most often Statecharts [14]. One of the most powerful and versatile of these tools is Rhapsody [17].

The fact that we concentrate on visual formalisms in this chapter does not mean that no other approaches exist or are feasible. There has been, for example, a tremendous amount of extremely interesting work done on using nonvisual approaches to modeling reactive systems, including temporal logic, process calculi, and process algebras. In many cases these approaches have also been used successfully in modeling biology; see, for example, [3, 22, 24, 25].

Another important point when we consider computational tools for modeling biology is that they must support ways to represent models and data that are natural and can become standard. User-friendliness, flexibility, and visuality are critical to integrating computational tools into experimental biology research. Biologists are not expected to become engineers. As a consequence, one of the challenges facing computer scientists is to design modeling languages and software tools that are more accessible to nonexperts.

4.3 On Statecharts

There are actually two approaches to modeling reactivity: intra-object and inter-object. In the realm of visual formalisms, they manifest themselves in the form of two families of languages: state based and scenario based [8, 18]. We will use Statecharts and Live Sequence Charts (LSCs) as representatives of these approaches. Both are visual languages with clear and rigorously defined syntax and semantics, and both facilitate the construction and execution of formal models. Statecharts specify the full state-based behaviors of each object in the system (e.g., a cell), whereas LSCs specify the multimodal (e.g., possible, necessary, and forbidden) scenarios that link together the objects by their possible sequences of behavior.

In this chapter we concentrate on Statecharts, and on the intra-object approach. The main reason is the fact that the use of states, or modes, appears to be a very natural way to view biological dynamics. In a state, part of the system (e.g., a cell) stays put, remains, or dwells for some non-zero amount of time; it moves on only as a result of the occurrence of some event (e.g., arrival of a signal, change in some relevant quantity or value).

Scenarios or pathways do not necessarily have an explicit notion of staying put, being, dwelling, residing, or other dynamic states.

Statecharts define behavior using a hierarchy of states with transitions, events, and conditions [14]. Classical finite-state machines (FSMs) and their state transition diagrams are extended by a semantically meaningful, hierarchical substating mechanism and by a notion of orthogonal simultaneity. Both of these approaches are reflected in the graphics themselves—the hierarchy by encapsulation and the orthogonality by adjacent portions separated by a dashed line. Orthogonal components can cooperate and know about each other by several means, including direct sensing of the state status in another component or by actions. The cooperation mechanism, within a single Statechart, has a broadcasting flavor. It helps to note that capsulated substates represent OR (actually this is XOR; exclusive or), whereas orthogonality is equivalent to AND.

Transitions in Statecharts become far more elaborate and rich than in conventional FSMs. They can start or stop at any level of the hierarchy, can cross levels, and in general can be hyperedges, because both sources and targets of transitions can contain sets of states. At any given point in time, a Statechart will be in a vector, or combination, of states, whose length is not fixed. Exiting and entering orthogonal components on the various levels of the hierarchy continuously changes the size of the state vector. Default states generalize start states, and they, too, can cross levels and be of hyperedge nature. Also, the language has history connectors, conditions, selection connectors, and more. A transition can be labeled with an event and optionally with a parenthesized condition, as well as with Mealy-like outputs, or actions. (Actions can also occur within states, as in the Moore style.)

Using Rhapsody (or other similar tools), a Statechart model can be compiled into executable reactive code (for example, in Java or C++). At run time, one state in each orthogonal component is active, which is the current state of this component. Thus the state of an object is identified by the set of active states of all of its components. As the simulation advances, various events move the active state in each orthogonal component from one state to another. Concurrent execution is naturally captured by the running of the model or, alternatively, in the executable code that is generated from it.

When using the object-oriented version of Statecharts (e.g., as supported by Rhapsody), the underlying structure of the objects gives them a measure of modularity and hierarchy. One can add or remove objects, connections, states, and transitions at any level. Statecharts also have ample means to specify rich kinds of behavior, such as concurrency, stochasticity, chain reactions, and time-dependent or time-constrained actions.

4.4 Statecharts for Biology

Since the mid-1980s, Harel [14] has often claimed that biological systems should be viewed as systems in the same way we know them in the world of computing and software engineering, and that biological modeling should be attempted using languages and tools constructed for developing reactive systems. As mentioned in Section 4.3, one obvious approach is to use states. State-based models define the behavior of objects (or other elements of the system under description) over time, based on the various states that an object can be in over its lifetime. In other words, states are abstract situations in an object's life cycle. Interacting state machines can specify causal relationships between state changes in different machines. These models describe how objects communicate and collaborate as well as how they behave under different circumstances. Interacting state machine models are particularly suitable for describing mechanistic models of biological systems that are well understood qualitatively. Such models do not require much quantitative data relating to the number of molecules and reaction rates. They allow the creation of abstract, high-level models and the application of strong analysis tools such as model checking.

Usually, the state of an object is determined in part by the states of its subobjects. For example, significant portions of the state of a cell would be determined by the states of the various genes and proteins making up that cell. Each gene or protein would then have its own reaction to the presence or absence of some other molecules, and the change in the state of the cell would be influenced greatly by the interdependent state changes of all parts. A hierarchical structure allows one to view a system at different levels of detail, and hierarchy in the behavioral description itself is no exception. Although many different languages are available in which to express interacting state machine models, Statecharts stand out—among other reasons—because of their support of hierarchy. In fact, a large part of the appeal of the language derives from its hierarchical description capabilities. The possibility of hierarchical structuring is extremely useful in cases where the behavior is distributed over many cells, and where multiple copies of the same process are executed in parallel.

Another advantage of Statecharts compared to other state-based formalisms, such as Reactive Modules [1], is the fact that they are visual. The user can draw states and state changes, and good supporting tools will be able to execute the resulting models, enabling relatively easy and intuitive programming even for nonspecialists. Using the visual language of Statecharts, Kam et al. [20] developed a model that described the various stages in the life span of a T cell, and the transitions between these stages. The initial T-cell model was followed by a more extensive animated model of T-cell differentiation in the thymus, carried out by Efroni et al. [9, 11].

That project introduced an idea called reactive animation (RA) [10], in which a reactive system (e.g., built using Statecharts) drives the display of animation software to visualize the model. At run time, the front end displays the simulation continuously and provides the means to interact with it. Initially, RA was implemented in an ad hoc fashion—that is, one reactive system engine to one animation tool. More recently, the technique has been upgraded, resulting in a generic platform that enables interaction between various tools such as multiple reactive engines, 3D animation, real-time analysis, and more [27].

Reactive systems, especially when used in modeling biology, call our attention to their emergent properties. These properties of the behavior of the system, taken as a whole, are not expressed by any one of the lower-scale components that constitute it. Life, for example, is an emergent property: None of the component molecules of a cell is alive; only a whole cell lives.

Following an argument made by Cohen and Harel [7], one notices that emergence is difficult to define in biological terms. It is a matter of scale. A cell emerges from the structured interactions of the molecules that form the cell. At the same time, the cell is orders of magnitude larger than its component molecular interactions. Obviously, you cannot see a cell from the inside at the molecular scale. The cell emerges only when you step back, zoom out, and look at the cellular system at a scale appropriate to seeing an entire cell. The cell emerges from its component interactions at the scale at which the cell functions as an object, with its own capabilities and its own interactions with other cells and molecules.

4.5 Actual Modeling Efforts

In some modeling projects carried out over the last few years, mainly by the Harel group at the Weizmann Institute of Science, we have illustrated the usefulness of Statecharts to reason about various biological processes. We briefly describe these model's in the remainder of this chapter.

4.6 Pancreatic Organogenesis in the Embryonic Mouse

The Statechart model of pancreatic organogenesis in the embryonic mouse [27] consists of a concurrent execution of pancreatic cells, which leads to the formation of the unique 3D structure of the organ (Figure 4.1). This model was studied by comparing simulations against relevant experimental data, and it reproduced some genetic ablation experiments in silico. As an example, Figure 4.2 shows a histological cut of the pancreas (left) and the emerging structure in the model at approximately the same day (middle). On the right, the figure shows the result of an in silico experiment in which

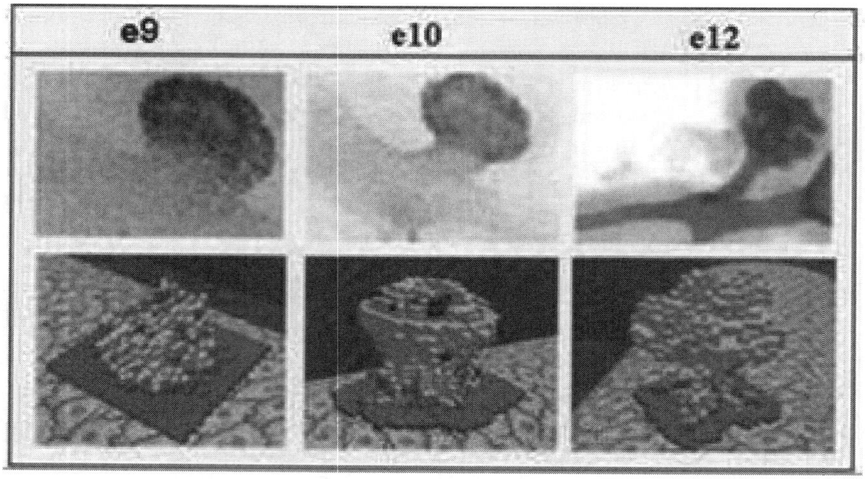

Figure 4.1 Pancreatic organogenesis in the embryonic mouse consists of a concurrent execution of pancreatic cells, which leads to the formation of the unique 3D structure of the organ [27]

the aorta was disabled, leading to a complete loss of structure. During the analysis of the pancreas model, concurrent execution of pancreatic cells gave rise to a property that corresponds well with endocrinic clusters that appear early in the developing organ in vivo [27].

Following the LSC model of vulval fate specification described by Kam et al. [21], Fisher et al. [12] created a Statecharts model of vulval fate specification based on the proposed mechanistic model of Sternberg and Horvitz from 1989 [28]. This work revealed that state-based mechanistic

Figure 4.2 A histological cut of the pancreas (left) and the emerging structure in the model at approximately the same day (middle), and the result of an in silico experiment, in which the aorta was disabled, leading to a complete loss of structure (right) [27]. Cell fate specification during *C. elegans* development

modeling is well suited to developmental genetics and can provide new insights into the temporal aspects of cell fate specification during *C. elegans* vulval development. More recent work [13] has been based on the more sophisticated understanding of vulval fate specification that we have today. Model checking allowed us to test the consistency of the current conceptual model for vulval precursor cell fate specification with an extensive set of observed behaviors and experimental perturbations of the vulval system. The analysis of this model predicted new genetic interactions between the signaling pathways involved in the patterning process, together with temporal constraints that may further elucidate the mechanisms underlying precise pattern formation during animal development. These predictions were also validated experimentally.

4.7 T-Cell Activation and Differentiation in the Thymus

The other models relate to the immune system and simulate the development and function of T cells in the thymus [9, 11]. Figure 4.3 shows analysis of the thymus model: microscopy images of knockout (A) and wild-type (B) thymus (left column), compared with the same area in the model (right column). On the right, Figure 4.3(a) shows an in silico experiment, in which the knockout of a single gene leads to a major change in the behavior of the cell population.

Analysis of the models also revealed several interesting emergent properties, which correspond well with biological phenomena. For example, the concurrent execution of T-cell development in the thymus led to the emergence of competitive behavior between the cells [11]. These properties were analyzed and studied, and were used to suggest some insights into the phenomena.

4.8 Development and Function of B Cells in the Lymph Node

Another line of investigation involved a Statechart and RA model of the lymph node; see the publication by Swerdlin et al. [29] for details. The effects of the amount of antigen as well as the actual size of the lymph node on the emergent properties of lymphocyte dynamics were studied. This model emphasized differentiation and anatomical localization. The dynamic organization of the lymph node visualized by RA sheds new light on how the immune system transforms antigen stimulation into a highly sensitive, yet buffered response.

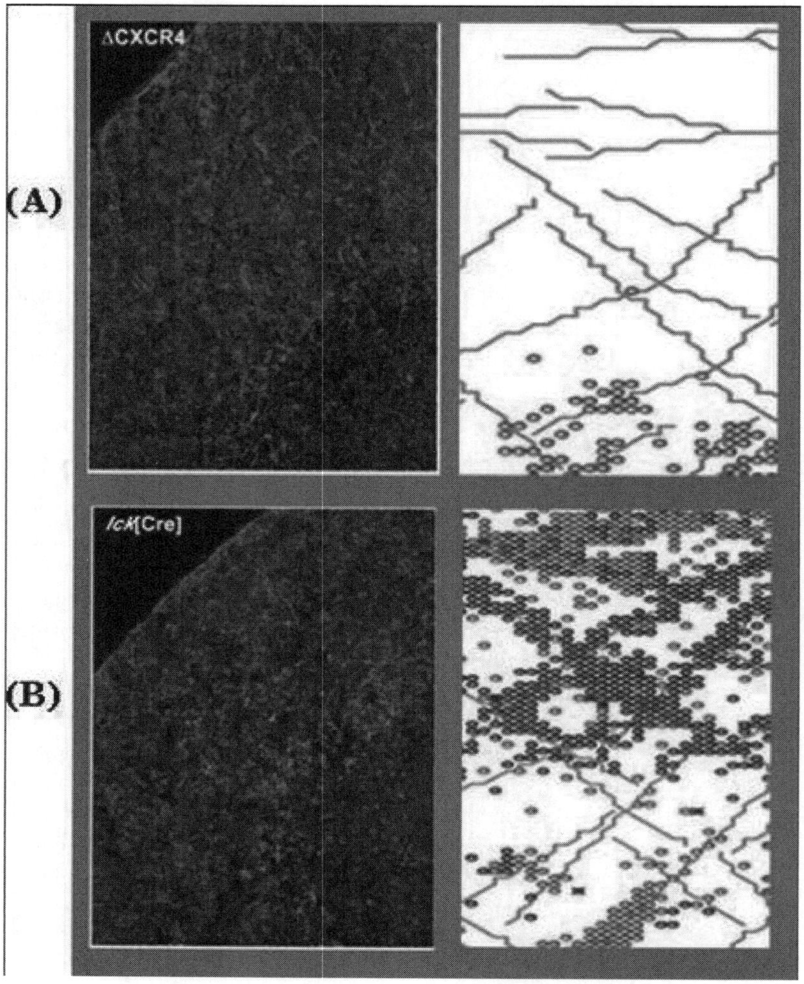

Figure 4.3 The thymus model [27]: microscopy images of knockout (A) and wild-type (B) thymus (left column [23]), compared with the same area in the model (right column)

4.9 Generic Cell Model

The most recent modeling work by the Harel group has resulted in a system called GemCell [2]. It contains a generic Statechart model of cell behavior, which captures the five main aspects of cell behavior: proliferation, death, movement, import, and export. This generic model is coupled with a database of biological specifics (DBS), which holds the information about the specific cellular system, and which is expected to be filled in by biologists with the relevant expertise. Modeling a particular segment of biology involves setting up the DBS to contain data about the specific behaviors and responses of the kinds of cells in the system under description. During execution, Statecharts read in the specific data and the combination runs just as described in the particular models.

EXERCISES

Construct a Statechart model for the Delta-Notch decision, as follows:

Consider a cell that is about to make a decision to adopt one of two fates: A or B. The decision is made according to the competition between two signaling pathways operating inside the cell and between neighboring cells. The following proteins are of interest: Notch receptor and its ligand Delta, which signify the level of the two signaling pathways.

Here is how the cell behaves. The Notch pathway encourages the cell to adopt fate A, while the Delta pathway encourages the cell to adopt fate B. In a normal cell, Notch starts in a low level, which gradually increases. When Notch reaches a certain level (threshold), it forces the cell to adopt fate A. At the same time, Delta encourages the Notch in the neighboring cells (via the Notch receptors on neighboring cells) and inhibits the Notch in the same cell. Delta starts on a low level; if it is not encouraged, it decreases until it disappears. Delta may be encouraged by the level of Notch in the neighboring cell (sensed by Delta). If Delta reaches a certain threshold, it inhibits the Notch and causes the cell to adopt fate B.

The following behaviors are observed:

1. When a cell is run in isolation, the Notch should prevail and the cell should assume fate A.
2. When two cells are run in parallel, either of them can assume fate A, in which case the other assumes fate B. There are rare cases where both cells assume fate A.
3. When one of the cells gets an external boost to the pathway, it is always the case that this cell adopts fate A and the other cell adopts fate B. Test your model on varying numbers of cells and see which configurations of fates (A and B) your model yields.

Acknowledgments

We would like to thank the many colleagues with whom we have worked on biological modeling, and especially those with whom the work was dominated by the use of Statecharts. Central among these are Irun R. Cohen, Sol Efroni, and Yaki Setty.

References

[1] R. Alur and T. A. Henzinger. Reactive modules. *Proceedings of the 11th Annual IEEE Symposium on Logic in Computer Science,* 207–218, 1996.

[2] H. Amir-Kroll, A. Sadot, I. R. Cohen, and D. Harel. GemCell: A generic platform for modeling multi-cellular biological systems. *Theoretical Computer Science*, 391(3):276–290, 2008.

[3] L. Cardelli. Brane calculi. *Proceedings of Computational Methods in Systems Biology*, 257–278, 2004.

[4] I. R. Cohen. *Tending Adam's Garden: Evolving the Cognitive Immune Self*. New York: Academic Press, 2000.

[5] I. R. Cohen. Informational landscapes in art, science and evolution. *Bulletin of Mathematics and Biology,* 68:1213–1229, 2006.

[6] I. R. Cohen, H. Atlan, and S. Elfroni. Genetics as explanation: Limits to the human genome project. In *Encyclopedia of Life Science (ELS)*, Chichester: John Wiley & Sons, Ltd., 2009.

[7] I. R. Cohen and D. Harel. Explaining a complex living system: Dynamics, multi-scaling and emergence. *Journal of Royal Society Interface,* 4:175–182, 2007.

[8] W. Damm and D. Harel. LSCs: Breathing life into message sequence charts. *Formal Methods in System Design,* 19(1):45–80, 2001.

[9] S. Efroni, D. Harel, and I. R. Cohen. Towards rigorous comprehension of biological complexity: Modeling, execution and visualization of thymic T cell maturation. *Genome Research,* 13:2485–2497, 2003.

[10] S. Efroni, D. Harel, and I. R. Cohen. Reactive animation: Realistic modeling of complex dynamic systems. *IEEE Computer,* 38(1):38–47, 2005.

[11] S. Efroni, D. Harel, and I. R. Cohen. Emergent dynamics of thymocyte development and lineage determination. *PLoS Computational Biology,* 3(1):127–136, 2007.

[12] J. Fisher, N. Piterman, E. J. A. Hubbard, M. J. Stern, and D. Harel. Computational insights into *C. elegans* vulval development. *Proceedings of National Academy of Science*, 102(6):1951–1956, 2005.

[13] J. Fisher, N. Piterman, A. Hajnal, and T. A. Henzinger. Predictive modeling of signaling crosstalk during *C. elegans* vulval development. *PLoS Computational Biology*, 3(5):e92, 2007.

[14] D. Harel. Statecharts: A visual formalism for complex systems. *Science of Computer Programming,* 8:231–274, 1987.

[15] D. Harel. On visual formalisms. *Communications of the Association Computing Machinery,* 31(5):514–530, 1988.

[16] D. Harel. A grand challenge for computing: Towards full reactive modeling of a multi-cellular animal. *Bulletin of the EATCS*, 81:226–235, 2003.

[17] D. Harel and E. Gery. Executable object modeling with Statecharts. *Computer*, 30(7):31–32, 1997.

[18] D. Harel and R. Marelly. *Come, Let's Play: Scenario-Based Programming Using LSCs and the Play-Engine*. Berlin: Springer-Verlag, 2003.

[19] D. Harel and A. Pnueli. On the development of reactive systems. In K. R. Apt, editor, *Logics and Models of Concurrent Systems*, NATO ASI Series, Vol. F-13, pages 477–498. Berlin: Springer-Verlag, 1985.

[20] N. Kam, I. R. Cohen, and D. Harel. The immune system as a reactive system: Modeling T cell activation with Statecharts. *Visual Languages and Formal Methods* (VLFM'01), *part of IEEE Symposium on Human-Centric Computing* (HCC'01), 15–22, 2001.

[21] N. Kam, D. Harel, H. Kugler, R. Marelly, A. Pnueli, E. J. A. Hubbard, and M. J. Stern. Formal modeling of *C. elegans* development: A scenario-based approach. *Proceedings of the Workshop on Computational Methods in Systems Biology (ICMSB), volume 2602 of Lecture Notes in Computer Science*, Pages 4–x. Berlin: Springer-Verlag, 2003. (Revised version in G. Ciobanu and G. Rozenberg, editors, *Modeling in Molecular Biology*, pages 151–173. Berlin: Springer, 2004.)

[22] M. Z. Kwiatkowska, G. Norman, D. Parker, O. Tymchyshyn, J. Heath, and E. Gaffney. Simulation and verification for computational modelling of signalling pathways. *Proceedings of the Winter Simulation Conference,* 1666–1674, 2006.

[23] J. Plotkin, S. E. Prockop, A. Lepique, and H. T. Petrie. Critical role for CXCR4 signaling in progenitor localization and T cell differentiation in the postnatal thymus. *Journal of Immunology,* 171:4521–4527, 2003.

[24] C. Priami and P. Quaglia. Modeling the dynamics of biosystems. *Briefings in Bioinformatics,* 5(3):259–269, 2004.

[25] A. Regev, W. Silverman, and E. Y. Shapiro. Representation and simulation of biochemical processes using the pi-calculus process algebra. *Proceedings of the Pacific Symposium on Biocomputing,* 459–470, 2001.

[26] A. Sadot, J. Fisher, D. Barak, Y. Admanit, M. J. Stern, E. J. A. Hubbard, and D. Harel. Towards verified biological models. *IEEE/ACM Transactions on Computing, Biology and Bioinformatics,* 5(2):1–12, 2008.

[27] Y. Setty, I. R. Cohen, Y. Dor, and D. Harel. Four-dimensional realistic modeling of pancreatic organogenesis. PNAS December 23, 2008, 105(51):20374–20379.

[28] P. W. Sternberg and H. R. Horvitz. The combined action of two intercellular signaling pathways specifies three cell fates during vulval induction in *C. elegans. Cell,* 58:679–693, 1989.

[29] N. Swerdlin, I. R. Cohen, and D. Harel, The lymph node B cell immune response: Dynamic analysis in-silico. *Proceedings of the IEEE* (Special issue on *Computational System Biology*), 96(8):1421–1443, 2008.

Chapter

5

A Visual Process Calculus for Biology

Andrew Phillips

Microsoft Research
7 JJ Thomson Avenue
CB3 0FB Cambridge, UK

Abstract This chapter presents a visual process calculus for designing and simulating computer models of biological systems. The calculus is based on a graphical variant of stochastic pi-calculus, extended with mobile compartments, and the simulation algorithm is based on standard kinetic theory of physical chemistry. The calculus forms the basis of a formal visual programming language for biology. The basic primitives of the calculus are first introduced by a series of examples involving genes and proteins. More complex features of the calculus are then illustrated by examples involving gene networks, cell differentiation, and immune system response. The main benefit of the calculus is its ability to model large systems incrementally, by directly composing simpler models of subsystems. The formal nature of the calculus also facilitates mathematical analysis of models, which in the future could help provide insight into some of the underlying properties of biological systems.

5.1 Introduction

In many respects, biological systems are like massively parallel, highly complex, error-prone computer systems. The genetic code of an entire organism is stored in digital form as DNA, but instead of a binary code there are four letters: G, A, T, and C. A specialized protein travels along the DNA to read the code, much like reading a sequence of instructions in a program. The code itself is divided into genes with different functions, in the same way that a computer program is divided into functional modules. The genes produce RNA, which is more accessible than the entire DNA sequence, like loading a module into memory. The RNA produces proteins, which fold into different shapes to perform the essential functions of a living organism. Some of these proteins can interact with the DNA to switch on different genes at different times, in the same way that some modules of a computer program can trigger the execution of other modules. In general, the genetic code can be viewed not as a single program, but rather as a *library* of programs that are executed on demand in response to environmental signals [22].

The Human Genome Project had the ambitious goal of mapping out the complete genetic code in humans. Scientists had hoped that by listing this code, they would be able to unravel the mysteries of how the human body functions. Instead, the code raised many more questions than it answered. It was then that the field of systems biology rose to the occasion, with the ultimate goal of being able to understand and predict the behavior of biological organisms. Systems biology can be broadly divided into two complementary approaches. On the one hand, scientists are doing experiments in the lab and studying the results, with the goal of inferring key properties of biological systems. On the other hand, they are using this knowledge to build detailed models of systems and then testing these models in the lab. Such models represent powerful tools, allowing scientists to simulate a range of experiments on a computer before they test the most promising ones, thereby saving resources and enabling more effective research. The models also help to clarify the mechanisms of how a biological system functions, and are beginning to play a role in understanding disease.

In recent years we have witnessed a rise in the publication of biological models. As these models grow in size, however, they are becoming increasingly difficult to understand, maintain, and extend. Many published models consist of hundreds of reactions, but as our knowledge of biological systems continues to increase, models consisting of tens of thousands of reactions will soon be commonplace. Clearly, writing a model of this size as a single list of reactions is not a scalable solution, for the same reasons that we would not write a large software program as a single list of thousands of instructions. Rather, we need a way of decomposing a large biological model into a collection of smaller, more manageable building blocks. Since the early days of computer programming, many high-level languages have successfully been developed to improve program modularity. However, biological programs differ from traditional computer programs in two fundamental ways: (1) they are massively parallel, and (2) each instruction has a certain probability of executing, so we are never quite sure what will happen next. These fundamental differences suggest a need for specialized programming languages for biology.

Over the years, there has been considerable research on programming languages for complex parallel computer systems. Some of this research is also applicable to biological systems, which are typically highly complex and massively parallel. One example of such a language is the pi-calculus [18, 33], which was originally proposed by Milner et al. [19] as a way to model communicating computer networks that can dynamically reconfigure themselves over time. A stochastic extension was later introduced by Priami et al. [28], to model the performance of computer networks in the presence of communication delays. More recently, the stochastic pi-calculus has been used to model and simulate a range of biological systems [17, 28], an approach pioneered by Regev et al. [30]. In addition, a graphical variant of the calculus has been defined [26], to help make the calculus more broadly accessible. The stochastic pi-calculus allows the components of a biological system to be modeled independently, rather than modeling the individual reactions, thereby enabling large models to be constructed by direct composition of simple components [1]. It also facilitates mathematical analysis of systems using a range of established techniques, which in the future could help provide insight into some of the underlying properties of biological systems. Various stochastic simulators have been developed for the calculus [25, 28], which are used to perform virtual experiments on biological models. Such in silico experiments can be used to formulate testable hypotheses on the behavior of biological systems, as a guide to future experimentation in vivo.

This chapter presents the SPiM calculus, a visual process calculus for designing and simulating computer models of biological systems. It is based on a graphical variant of the stochastic pi-calculus, originally presented in [26], extended with mobile compartments. The basic primitives of the calculus are introduced through a series of examples involving genes and proteins. More complex features of the calculus are then illustrated by examples involving gene networks, cell differentiation, and immune system response. All of the biological models in this chapter were simulated using the SPiM simulator [25] and are available for download online, together with the simulator itself.[1]

5.1.1 SPiM Processes and Chemical Reactions

The Gillespie algorithm [13] was introduced in the 1970s for the exact stochastic simulation of coupled chemical reactions. Since then, a vast

[1] http://research.microsoft.com/spim

number of chemical and biological models have been simulated using this approach. Even today, many biological models continue to be published as collections of chemical reactions, and the format is supported by most modeling standards such as the Systems Biology Markup Language [15].

Essentially, the SPiM calculus allows a system of chemical reactions to be rewritten as a collection of modular processes, which interact with one another by means of passing messages. Figure 5.1 presents a comparison between chemical reactions and SPiM processes for a simple model of protein activation. For both models, the graphical representation at the top is equivalent to the textual representation at the bottom. The chemical reaction model of Figure 5.1(a) is constructed by listing the individual *reactions* in the system. An active protein X_p can interact with a protein Y through an a reaction, to produce a protein X and an active protein Y_p. The reverse reaction d can also occur. For the graphical representation, each shape represents a protein in a particular state and each box represents a reaction, with inbound edges (arcs) from reactants and outbound edges to products. In contrast, the SPiM model of Figure 5.1(b) is constructed by describing the behavior of the individual *components* in the system. This is achieved by splitting the reaction a into two complementary actions, a send $\overline{a}$ and a receive $\underline{a}$, and similarly for the reaction d. Thus an active protein X_p can send on $\overline{a}$ and evolve to a protein X, which can receive on $\underline{d}$ and evolve to X_p. Similarly, a protein Y can receive on $\underline{a}$ and evolve to an active protein Y_p, which can send on $\overline{d}$ and evolve to Y. For the graphical representation, each shape represents a protein in a particular state and each connected graph represents the set of possible states of a given protein, where a labeled edge represents an action that the protein can perform to change from one state to another. Because X_p can send on $\overline{a}$ and Y can receive on $\underline{a}$, the two proteins can interact with each other and evolve to a new state simultaneously.

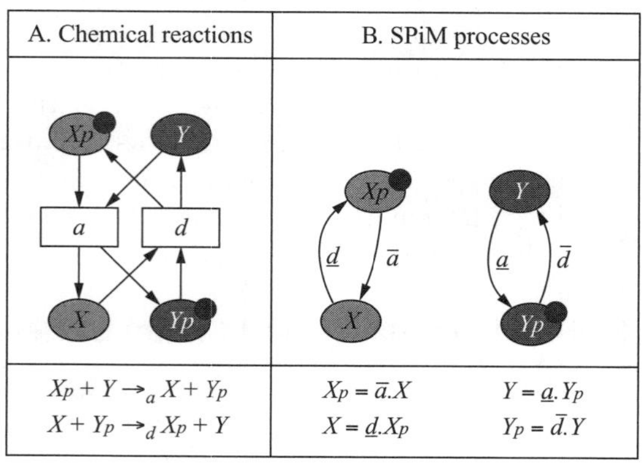

Figure 5.1 Comparison between chemical reactions (a) and SPiM processes (b) for a model of protein activation

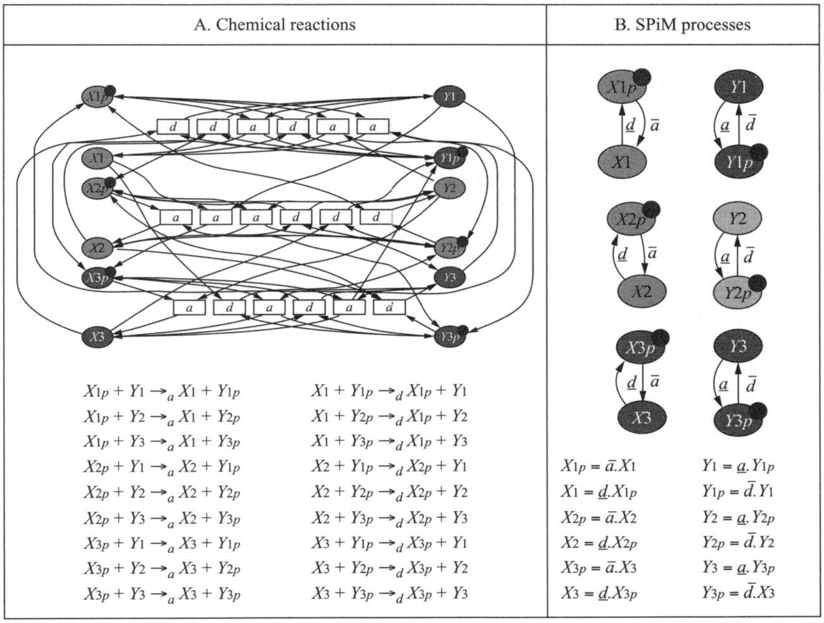

Figure 5.2 Comparison between chemical reactions and SPiM processes for a combinatorial model of protein activation. (a) For the chemical reaction model, there are 18 reactions, 9 forward reactions between X_{ip} and Y_j and 9 reverse reactions between X_i and Y_{jp} with $i, j \in \{1, \ldots, 3\}$. In the general case of $i \in \{1, \ldots, N\}$ and $j \in \{1, \ldots, M\}$, there will be $N \cdot M$ forward reactions and $N \cdot M$ reverse reactions. (b) For the SPiM model, there are 12 actions, 6 forward actions for X_{ip}, Y_j and 6 reverse actions for X_i, Y_{jp} with $i, j \in \{1, \ldots, 3\}$. In the general case of $i \in \{1, \ldots, N\}$ and $j \in \{1, \ldots, M\}$, there will be $N + M$ forward actions and $N + M$ reverse actions

The SPiM model explicitly represents the fact that the X_p protein evolves to X and the Y protein evolves to Y_p after the interaction takes place. This contrasts with the chemical reaction model, which does not explicitly state which product comes from which reactant. Although we can guess by looking at the reactant and product names, we could equally well interpret the reactions to mean that X_p becomes Y_p and Y becomes X.

Another feature of the SPiM model is that the description of the X protein can be given independently of that of the Y protein, allowing the behaviors of the two proteins to be modeled separately. A notion of complementary send $\overline{a}$ and receive $\underline{a}$ is needed to decompose the system into individual modules in this way. If we simply wrote $X_p = a.X$ and $Y = a.Y_p$ and allowed all the a actions to interact, there would be nothing to prevent two X_p proteins from interacting with each other, which should not be possible. The use of complementary actions in the model is also consistent with the physical characteristics of the system, because the protein X_p

actually sends its phosphate group p to a receiver protein Y, which then becomes phosphorylated as Y_p. Conversely, the phosphorylated protein Y_p sends its phosphate group p back to protein X. More generally, complementary actions can be used to represent a range of biological interactions, such as two proteins with complementary binding sites, two RNA strands with complementary sequences, two atoms with complementary electron configurations, and so forth.

In addition to improving modularity, the SPiM calculus can help to reduce the size of a biological model. Consider the models in Figure 5.2, which describe a collection of proteins that activate one another. For both models, the graphical representation at the top is equivalent to the textual representation at the bottom. Each of the proteins $X_{1p}, \ldots, X_{3p}$ can activate each of the proteins $Y_1, \ldots, Y_3$, and similarly for the proteins $Y_{1p}, \ldots, Y_{3p}$ and $X_1, \ldots, X_3$. For the chemical reaction model in Figure 5.2(a), each interaction needs to be described explicitly, resulting in a combinatorial explosion in the number of reactions. For the SPiM model in Figure 5.2(b), instead of explicitly stating which proteins interact with which other proteins, it is necessary to describe only the channels on which each protein can send and receive. Implicitly, all senders on a given channel can interact with all receivers on the same channel, resulting in a more compact model whose size varies linearly with the number of proteins. As the system grows in size, the advantages of a modular representation become more noticeable, as the behaviors of individual proteins can be modified without changing the rest of the model.

5.2 Basic Examples

This section presents the basic constructs of the SPiM calculus, through a collection of simple examples. Although these examples are presented at the level of interacting genes and proteins, the constructs of the calculus can be used equally well at the level of interacting cells or organisms. The simulation results for the examples were obtained using the SPiM simulator, which executes a SPiM model according to an exact simulation algorithm [25] based on standard principles of physical chemistry [13].

5.2.1 Protein Production

The first example models the production of protein P from a gene G, as shown in Figure 5.3. In the SPiM model of Figure 5.3(a), the graphical representation at the top is equivalent to the textual representation at the bottom. The gene G can do a *produce* action, after which a new protein P is executed in parallel with the gene, written $(G \mid P)$. Graphically, the parallel execution is represented as a box with outbound edges to

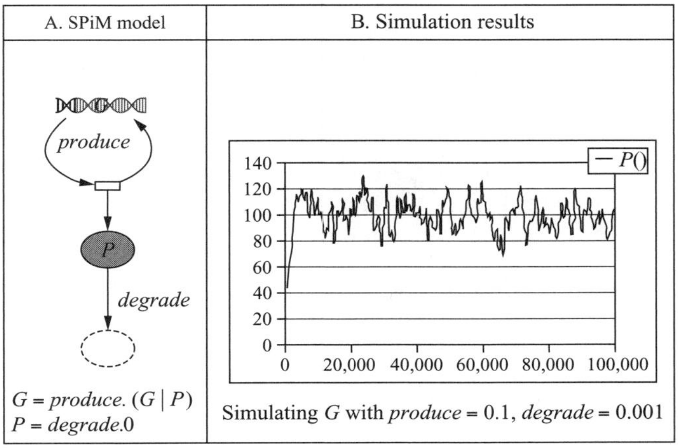

Figure 5.3 A SPiM model of protein production (a), with associated simulation results (b)

G and P. The protein P can then do a *degrade* action, after which the empty process is executed, written $\mathbf{0}$. Graphically, the empty process is represented as a node with a dotted outline. To obtain the simulation results of Figure 5.3(b), the *produce* and *degrade* actions are associated with corresponding rates given by 0.1 and 0.001, respectively. Each rate characterizes an exponential distribution, such that the probability of an action with rate r occurring within time t is given by $F(t) = 1 - e^{-rt}$. The average duration of the action is given by the mean $1/r$ of this distribution. Thus the *produce* action will take, on average, 10 time units, while the degrade action will take, on average, 1000 time units. In this example the time units are assumed to be seconds, but in general they can be any unit that is appropriate for the system under consideration. The simulation results show the population of protein P over time. The population of the gene G is always 1 and is not shown. Initially there is one gene G and no protein P in the system. The number of proteins increases over time until it reaches an equilibrium of about 100, and then continues to fluctuate around this equilibrium.

The simulation of the model in Figure 5.3 is described in more detail in Figures 5.4 and 5.5. The execution steps in Figure 5.4 describe the evolution of the system over time, where the graphical representation at the top is equivalent to the textual representation at the bottom. Each frame represents a state of the system, where the populations of the gene G and protein P are indicated next to the corresponding nodes in the graph, and the next actions to be executed are highlighted in gray.

The calculations in Figure 5.5 show the *propensities* of the reactions at each step of the simulation. For the system of Figure 5.3, there are

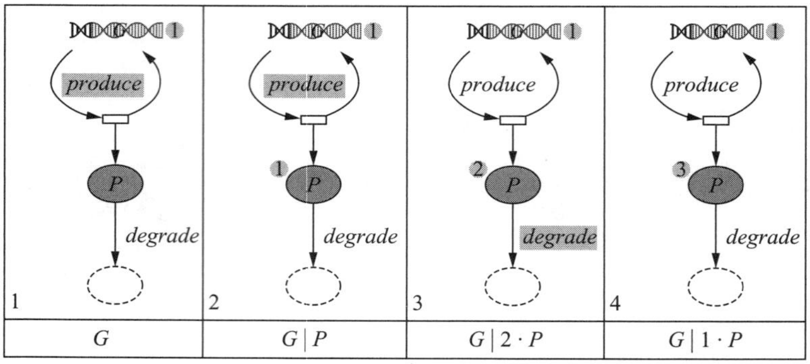

Figure 5.4 Execution steps for the protein production model of Figure 5.3. (1) Initially there is one gene G and no protein P. The gene then does a *produce* action. (2) A new protein is created in parallel with the gene, which then does a second *produce* action. (3) A second protein is created, and one of the proteins then does a *degrade* action. (4) One of the proteins is degraded

two possible reactions: *produce* and *degrade*. The propensity $\mathcal{R}(produce)$ of the *produce* reaction is given by the number of genes G that can perform this action, written $[G]$, multiplied by the rate 0.1 of the action. Similarly, the propensity $\mathcal{R}(degrade)$ of the *degrade* reaction is given by the number of proteins P that can perform this action, written $[P]$, multiplied by the rate 0.001 of the action. In accordance with [13], the probability of executing a given reaction is proportional to its propensity. To calculate

#	$R(produce) = [G] \cdot produce$	$R(degrade) = [P] \cdot degrade$
1	0.1 = 1·0.1	0 = 0·0.001
2	0.1 = 1·0.1	0.001 = 1·0.001
3	0.1 = 1·0.1	0.002 = 2·0.001
4	0.1 = 1·0.1	0.001 = 1·0.001
...	0.1 = 1·0.1	0.1 = 100·0.001

Figure 5.5 Propensity calculations for the execution steps of Figure 5.4, where the probability of a reaction is proportional to its propensity. (1) The system starts with a single gene G, which can produce a protein P with propensity 0.1. (2) A second protein can be produced with propensity 0.1. (3) One of the proteins can be degraded with propensity 0.002. Although the propensity of the *degrade* reaction is much lower than that of the *produce* reaction, there is still a small chance that a protein can degrade before another is produced. (...) Eventually an equilibrium is reached at approximately 100 proteins, where the propensity $[G]\cdot 0.1$ of the *produce* reaction is equal to the propensity $[P]\cdot 0.001$ of the *degrade* reaction

the probability of each reaction at a given instant, the simulation algorithm adds up the propensities of all possible reactions in the system and divides the propensity of each reaction by this total. The system reaches an equilibrium when the propensity of the *produce* reaction is equal to the propensity of the *degrade* reaction.

5.2.2 Protein Interaction

The second example models the interaction between proteins, as shown in Figure 5.6, where the SPiM model of Figure 5.6(a) is the same as in Figure 5.1(a). An active protein X_p can send on $\overline{a}$ and evolve to a protein X, which can receive on $\underline{d}$ and evolve to X_p. Similarly, a protein Y can receive on $\underline{a}$ and evolve to an active protein Y_p, which can send on $\overline{d}$ and evolve to Y.

To obtain the simulation results of Figure 5.6(b), the channels a and d are associated with corresponding rates given by $\rho(a) = 100$ and $\rho(d) = 10$, respectively, where the rate of the channel denotes the rate of a single interaction on this channel. As in the previous example, each rate r characterizes an exponential distribution with mean $1/r$. Thus an interaction on channel a will take, on average, 0.01 time units, while an interaction on channel d will take, on average, 0.1 time units. The simulation results show the populations of proteins X_p and X over time. The populations of Y and Y_p are not shown because they are equal to those of X_p and X, respectively. Initially there are $100 \cdot X_p$ and $100 \cdot Y$ proteins. The number of X_p proteins decreases and the number of X proteins increases over time until the system reaches an equilibrium of about $24 \cdot X_p$ and $76 \cdot X$ proteins, and the populations then continue to fluctuate around this equilibrium.

The simulation of the model in Figure 5.6 is described in more detail in Figures 5.7 and 5.8. The execution steps in Figure 5.7 describe the

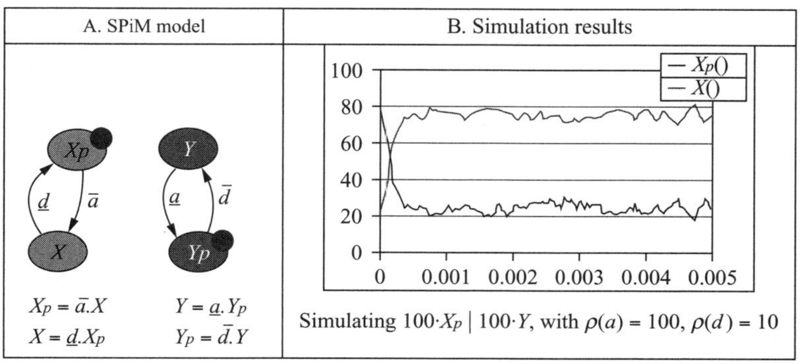

Figure 5.6 A SPiM model of protein interaction (a), with associated simulation results (b)

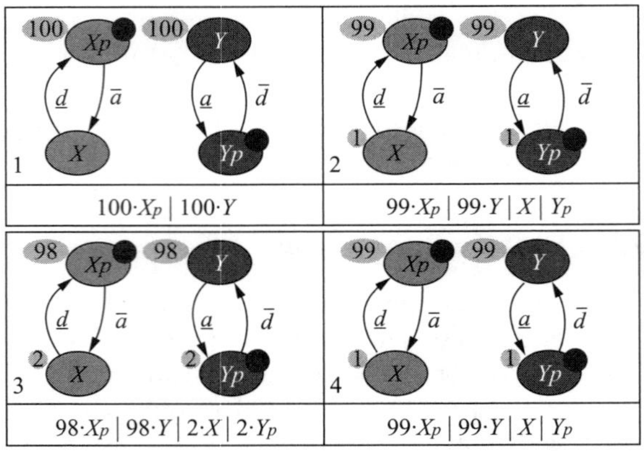

Figure 5.7 A sequence of execution steps for the protein interaction model of Figure 5.6. (1) Initially there are $100 \cdot X_p$ and $100 \cdot Y$ proteins. An X_p and a Y protein interact on channel a. (2) The two interacting proteins are converted to X and Y_p, respectively. An additional X_p and Y protein interact on channel a. (3) An X and Y_p protein interact on channel d. (4) The two interacting proteins are converted back to X_p and Y, respectively

evolution of the system over time, and the calculations in Figure 5.8 show the propensities of each reaction at each step of the simulation. There are two possible reactions in the system: a and d. The propensity $\mathcal{R}(a)$ of the a reaction is given by the number of possible interactions on channel a times the rate of a single interaction. Because each protein X_p can interact with each protein Y on channel a, the number of possible interactions on

#	$R(a)$	=	$[Xp] \cdot [Y] \cdot \rho(a)$	$R(d)$	=	$[X] \cdot [Yp] \cdot \rho(d)$
1	1,000,000	=	100·100·100	0	=	0·0·10
2	980,100	=	99·99·100	10	=	1·1·10
3	960,400	=	98·98·100	40	=	2·2·10
4	980,100	=	99·99·100	10	=	1·1·10
...	57,600	=	24·24·100	57,760	=	76·76·10

Figure 5.8 Propensity calculations for the execution steps of Figure 5.7, where the probability of a reaction is proportional to its propensity. (1) The system starts with $[X_p] = [Y] = 100$. An X_p and a Y protein can interact on channel a with propensity 1,000,000. (2) An additional X_p and Y protein can interact with propensity 980,100. (3) An X and Y_p protein can then interact on channel d with propensity 40. (...) Eventually an equilibrium is reached at around $[X_p] = [Y] = 24$ and $[X] = [Y_p] = 76$, where the propensity $[X_p] \cdot [Y] \cdot 100$ of the a reaction is roughly equal to the propensity $[X] \cdot [Y_p] \cdot 10$ of the d reaction

a is given by the population of X_p, written $[X_p]$, times the population of Y, written $[Y]$. The rate of a single interaction is given by the rate of the channel a, written $\rho(a)$. Therefore, the propensity $\mathcal{R}(a)$ of the a reaction is given by $[X_p] \cdot [Y] \cdot \rho(a)$. Similarly, the propensity $\mathcal{R}(d)$ of the d reaction is given by $[X] \cdot [Y_p] \cdot \rho(d)$. The system reaches an equilibrium when the propensity of the a reaction is equal to the propensity of the d reaction.

5.2.3 Protein Binding

The third example models the binding and unbinding of proteins, as shown in Figure 5.9. In the SPiM model of Figure 5.9(a), the graphical representation at the top is equivalent to the textual representation at the bottom. To model binding and unbinding reactions, we associate a given channel b with corresponding binding and unbinding channels, written ^{+}b and ^{-}b, respectively. A protein X can send and bind on $^{+}\overline{b}$ and evolve to a protein X', which can send and unbind on $^{-}\overline{b}$ and evolve to X. Similarly, a protein Y can receive and bind on $^{+}\underline{b}$ and evolve to a protein Y', which can receive and unbind on $^{-}\underline{b}$ and evolve to Y. The send and receive actions on binding channel ^{+}b represent the fact that the interacting proteins become bound to each other after the interaction takes place. The bound proteins can then unbind from each other by interacting on unbinding channel ^{-}b.

To obtain the simulation results of Figure 5.9(b), the channels ^{+}b and ^{-}b are associated with corresponding rates given by $\rho(^{+}b) = 100$ and $\rho(^{-}b) = 10$, respectively. The simulation results show the populations of proteins X and X' over time. The populations of Y and Y' are not shown because they are equal to those of X and X', respectively. Initially there are $100 \cdot X$

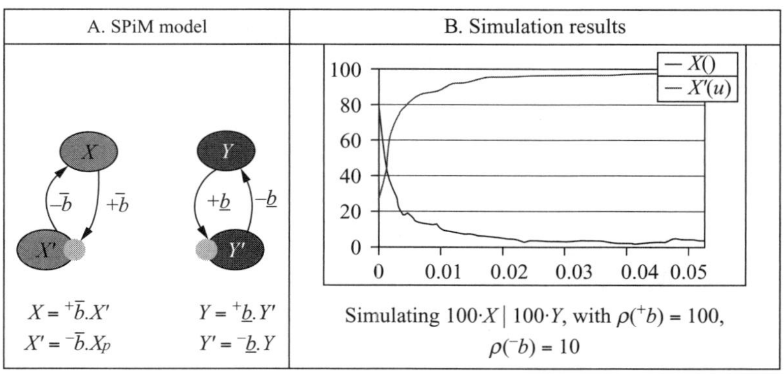

Figure 5.9 A SPiM model of protein binding (a), with associated simulation results (b)

and $100 \cdot Y$ proteins. The number of X proteins decreases and the number of X' proteins increases over time until the system reaches an equilibrium consisting of approximately $3 \cdot X$ proteins and $97 \cdot X'$ proteins.

The simulation of the model in Figure 5.9 is described in more detail in Figures 5.10 and 5.11. There are two possible reactions in the system: ${}^{+}b$ and ${}^{-}b$. As with the previous example, the propensity $\mathcal{R}({}^{+}b)$ of the ${}^{+}b$ reaction is given by the number of possible interactions on binding channel ${}^{+}b$ times the rate of a single interaction, written $[X] \cdot [Y] \cdot \rho({}^{+}b)$. After the interaction takes place, each X' protein is bound to a single Y' protein on a local channel ${}^{-}b$, and the pair of bound proteins is written $({}^{-}b)\,(X' \mid Y')$. The number of possible interactions on unbinding channel ${}^{-}b$ is given by the number of pairs of bound X' and Y' proteins, written $[({}^{-}b)\,(X' \mid Y')]$. Therefore, the propensity $\mathcal{R}({}^{-}b)$ of the ${}^{-}b$ reaction is given by $[({}^{-}b)\,(X' \mid Y')] \cdot \rho({}^{-}b)$. If we instead used a nonbinding interaction on channel b, then every X' protein would be able to interact with every other Y' protein, and the equilibrium would be 24 X' and Y' proteins, similar to the previous example. Because each X' protein can interact with only a single Y' protein, however, a reaction between X' and Y' is much less likely to occur, and the equilibrium is 97 X' and Y' proteins. This example illustrates how the use of binding interactions can significantly alter the equilibrium of a system.

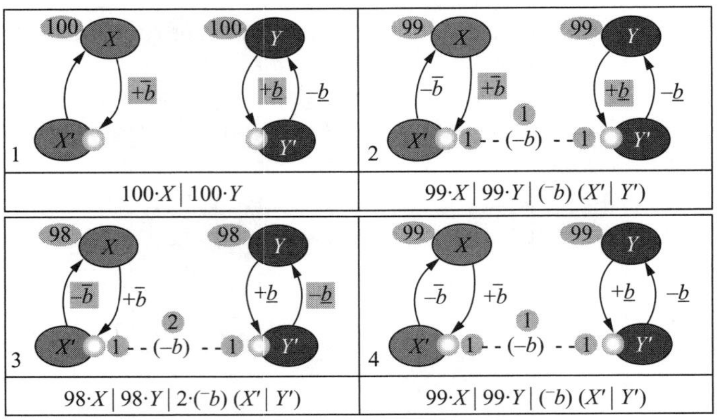

Figure 5.10 A sequence of execution steps for the protein-binding model of Figure 5.9. (1) Initially there are $100 \cdot X$ and $100 \cdot Y$ proteins. An X and a Y protein bind to each other by interacting on channel ${}^{+}b$. (2) The interacting proteins are converted to X' and Y', respectively, and are bound to form a complex $({}^{-}b)\,(X' \mid Y')$. An additional X' and Y' protein bind to each other by interacting on channel ${}^{+}b$, resulting in two complexes: $2 \cdot ({}^{-}b)\,(X' \mid Y')$. (3) An X' and a Y' protein in one of the complexes unbind from each other by interacting on the local channel ${}^{-}b$. (4) The two interacting proteins are converted back to X and Y, respectively

#	$R(^+b)$	$= [X] \cdot [Y] \cdot \rho(^+b)$	$R(^-b)$	$= [(^-b)\,(X' \mid Y')] \cdot \rho(^-b)$
1	1,000,000 =	100·100·100	0 =	0·10
2	980,100 =	99·99·100	10 =	1·10
3	960,400 =	98·98·100	20 =	2·10
4	980,100 =	99·99·100	10 =	1·10
...	900 =	3·3·100	970 =	97·10

Figure 5.11 Propensity calculations for the execution steps of Figure 5.10, where the probability of a reaction is proportional to its propensity. (1) The system starts with $[X] = [Y] = 100$. An X and a Y protein can bind by interacting on channel ^+b with propensity 1,000,000. (2) An additional X and Y protein can interact on ^+b with propensity 980,100. (3) A pair of bound X' and Y' proteins can unbind by interacting on ^-b with propensity 20. Eventually an equilibrium is reached at around $[X] = [Y] = 3$ and $[(^-b)\,(X' \mid Y')] = 97$, where the propensity $[X] \cdot [Y] \cdot 100$ of the ^+b reaction is roughly equal to the propensity $[(^-b)\,(X' \mid Y')] \cdot 10$ of the ^-b reaction

5.2.4 Summary

This section introduced the basic primitives of the SPiM calculus, using examples of protein production, protein interaction, and protein binding. The examples were presented in both graphical and textual forms, which could be used interchangeably.

The first example showed how a gene G could *produce* a new protein P in parallel with itself, written $G = produce.(G \mid P)$, and how a protein P could *degrade* to nothing, written $P = degrade.\mathbf{0}$. Both of these definitions gave the *behaviors* of the gene G and protein P. To run a simulation, initial *populations* were specified for processes G and P, and stochastic *rates* were associated with the *produce* and *degrade* actions, where the average duration of an action was defined as one over its rate. The *propensity* of each action was defined as the number of processes that can perform this action times the rate of the action. At each step of the simulation, a given action was chosen with probability proportional to its propensity. An equilibrium was reached when the propensities of both actions were roughly the same.

The second example showed how a protein X_p could send on $\overline{a}$ and evolve to a protein X, written $X_p = \overline{a}.X$, and how a protein Y could receive on $\underline{a}$ and evolve to a protein Y_p, written $Y = \underline{a}.Y_p$. Because these two proteins could send and receive on the same channel, they could interact on this channel and evolve simultaneously. The reverse interaction could also occur on channel d, where $X = \underline{d}.X_p$ and $Y_p = \overline{d}.Y$. Initial populations were specified for processes X and Y_p, and stochastic rates were associated with the channels a and d, where the average duration of a channel interaction was defined as one over the channel rate. The propensity of each channel interaction was defined as the number of sends on the channel times the number of receives on the channel times the rate of the channel. At each step of the simulation, a given channel interaction was chosen with probability proportional to its propensity. An equilibrium was reached when the propensities of both channel interactions were roughly the same.

The third example introduced a notion of binding and unbinding channels, written as ^{+}b and ^{-}b, respectively. The example showed how a protein X could send and bind on $^{+}\overline{b}$ and evolve to a protein X', written $X = {}^{+}\overline{b}.X'$, and how a protein Y could receive and bind on $^{+}\underline{b}$ and evolve to a protein Y', written $Y = {}^{+}\underline{b}.Y'$. After the binding interaction on channel ^{+}b, the two proteins became bound to each other, written $({}^{-}b)\,(X' \mid Y')$, and the bound proteins could then unbind by interacting on channel ^{-}b, where $X' = {}^{-}\overline{b}.X$ and $Y' = {}^{-}\underline{b}.Y$. Initial populations were specified for processes X and Y, and stochastic rates were associated to channels ^{+}b and ^{-}b. The key difference from the previous example was that each X' protein, became bound to a single Y' protein on channel ^{-}b. The propensity of an unbinding channel interaction was defined as the number of pairs bound by the channel times the rate of the channel. The example illustrated how the use of binding interactions could alter the equilibrium of a system.

These three examples illustrate some of the basic constructs of the SPiM calculus. More complex features of the calculus are illustrated in the following sections, and the full definition of the calculus is given in Section 5.7.

5.3 Gene Networks

5.3.1 Repressilator Model

One of the ultimate challenges of systems biology is to understand in detail how living systems function. Once this goal has been achieved, in principle we should be able to engineer living systems to behave in predictable ways. Such engineering of biological systems could also pave the way for new medical therapies and treatments.

In recent years, scientists have been investigating how to genetically engineer simple organisms such as bacteria, by inserting combinations of genes that interact with each other in predefined ways. For example, in one well-known experiment [11], *E. coli* were genetically engineered to repeatedly glow on and off. A network of three genes was inserted into the bacteria, consisting of the genes *tet*, *lambda*, and *lac*, which mutually repressed one another. More precisely, the *tet* gene produced proteins to block the *lambda* gene, which produced proteins to block the *lac* gene, which produced proteins to block the *tet* gene, completing the cycle. An additional *gfp* gene was inserted to produce a green fluorescent protein (GFP), which made the bacteria glow. The network was constructed so that the *tet* gene produced proteins that also blocked the *gfp* gene, thereby preventing the production of GFP. When engineered in live bacteria, the gene network caused the bacteria to repeatedly glow on and off. The resulting network was called the *Repressilator*, reflecting the fact that the mutual repression of genes gave rise to oscillations in GFP levels. The gene network was also reproduced in the descendants of the bacteria, which themselves continued to glow on and off.

The Repressilator network can be simulated in SPiM by first constructing a parameterized model of a gene with negative control as shown in Figure 5.12, based on previous work presented in [1] and [2]. The model is similar to the protein production model of Figure 5.3, except that the gene and the protein are parameterized by the channels on which they can interact. The gene G is parameterized by channels a, b, while the protein P is parameterized by channel b. The parameterized gene $G(a, b)$ can perform one of two actions: (1) it can *produce*, after which a new protein $P(b)$ is executed in parallel with the gene, or (2) it can receive on $\underline{a}$ and become blocked to $B(a, b)$. The choice between production and blocking is indicated using a (+) symbol. Graphically, this situation is represented by two outbound edges from node $G(a, b)$, labeled with the *produce* and $\underline{a}$ actions, respectively. If the *produce* action is chosen, then a new protein is produced in parallel with the gene. If the $\underline{a}$ action is chosen, then the

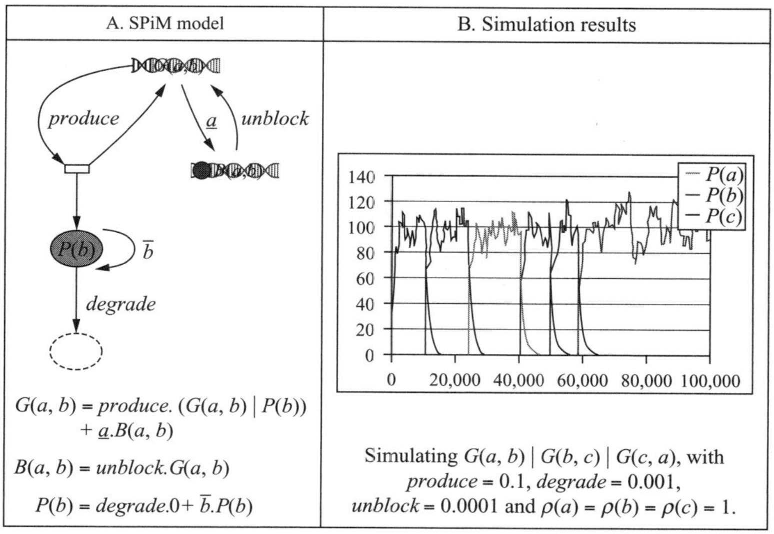

Figure 5.12 A SPiM model of a parameterized gene with negative control (a), with associated simulation results (b)

gene becomes blocked to $B(a, b)$ and is no longer able to produce. The blocked gene can then *unblock* to become active again. The parameterized protein $P(b)$ can also perform one of two actions: (1) it can *degrade*, after which the empty process is executed, or (2) it can send on $\overline{b}$ and continue executing. Therefore, $G(a, b)$ denotes a gene that is blocked by receiving on $\underline{a}$, and that produces proteins that in turn send on $\overline{b}$.

The use of complementary actions in the model is consistent with the physical characteristics of the system, as $G(a, b)$ denotes a gene with promoter region $\underline{a}$, which produces proteins $P(b)$ with binding site $\overline{b}$. When the promoter region is free, the gene is able to repeatedly produce proteins. When the promoter region $\underline{a}$ is blocked by a protein with a complementary binding site $\overline{a}$, the gene is no longer able to produce. The parameterized gene can be used to construct networks of arbitrary complexity, as illustrated in Figure 5.13. An auto-inhibitory gene can be defined as $G(b, b)$—that is, as a gene with promoter region $\underline{b}$ that produces proteins with binding site $\overline{b}$. Thus the proteins produced by the gene will, in turn, cause the gene to block. In this setting, the parameters (a, b) of the gene are replaced with actual channels (b, b), resulting in a gene that blocks itself. A bi-stable network can be defined as $G(a, b) \mid G(b, a)$—in other words, two genes that block each other. The first gene uses channels (a, b) as usual, while the second gene uses channels (b, a). Similarly, an oscillator network can be defined as $G(a, b) \mid G(b, c) \mid G(c, a)$.

The simulation results for the oscillator network of genes are shown in Figure 5.12(b), where the *produce*, *degrade*, and *unblock* actions are associated with corresponding rates given by 0.1, 0.001, and 0.0001, respectively, and the channels a, b, c are associated with rate 1. The results show the populations of proteins $P(a)$, $P(b)$, $P(c)$ over time. Initially, approximately 100 $P(a)$ proteins are produced and the other two proteins are absent. Subsequently, approximately 100 $P(b)$ proteins are produced, which repress the production of $P(a)$. Eventually, approximately 100 $P(c)$ proteins are produced, which repress the production of $P(b)$, completing the cycle. The alternate production of proteins continues in this order until the end of the simulation.

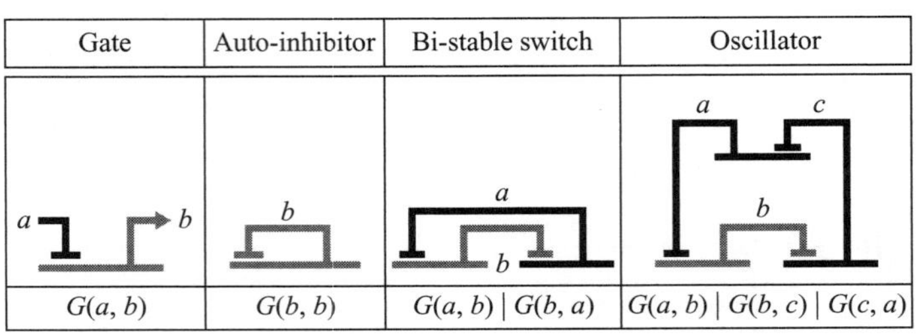

Figure 5.13 Modular construction of gene networks using the parameterized gene of Figure 5.12

From the simulation results, it is not immediately clear how the oscillations are initialized, or why they proceed in a specific order. To understand the mechanism by which the gene network produces these oscillations, it is necessary to closely examine the interactions between genes and proteins over time.

5.3.2 Model Execution

The simulation of the model in Figure 5.12 is described in more detail in Figures 5.14 and 5.15. Figure 5.14 shows the first few execution steps for the simulation results of Figure 5.12, where the definitions of $G(a,b)$, $G(b,c)$, and $G(c,a)$ are expanded so that the behavior of each gene is represented as a separate graph. The graph for each gene is simply a copy of the model in Figure 5.12, but with parameters (a,b) being replaced by (a,b), (b,c), and (c,a), respectively. Figure 5.15 shows the propensities of the different reactions for each of the execution steps of Figure 5.14. The propensities demonstrate how the first protein cycle is produced by a sequence of high-probability reactions.

There are three possible delay reactions in the system: *produce*, *degrade*, and *unblock*. The propensity of the *produce* reaction is given by the sum of the production propensities of each of the genes. The production propensity of gene $G(a,b)$ is given by the population of the gene times the rate of production, written $[G(a,b)] \cdot produce$, and similarly for genes $G(b,c)$ and $G(c,a)$. Thus the propensity of the *produce* reaction is given by $[G(a,b)] \cdot produce + [G(b,c)] \cdot produce + [G(c,a)] \cdot produce$. The choice of a particular gene to produce happens in two stages: (1) the *produce* reaction is chosen, and (2) one of the active genes is chosen to produce with equal probability. The propensities of the *unblock* and *degrade* reactions are defined in a similar fashion.

There are three possible interactions in the system, on channels a, b, and c. The propensity of an interaction on channel a is given by the number of proteins $P(a)$ times the number of genes $G(a,b)$ times the rate of the channel, written $[P(a)] \cdot [G(a,b)] \cdot \rho(a)$. This essentially determines the propensity of gene $G(a,b)$ to be blocked by protein $P(a)$. The greater the population of repressor proteins, the greater the blocking propensity. The propensities of the interactions on channels b and c are defined in a similar fashion.

When examining the simulation results of Figure 5.12 or the cartoon strip of Figure 5.14, it is difficult to understand the causal chain of events leading up to the first protein cycle. To clarify the interactions that take place, together with the consequences of these interactions, Figure 5.16 represents the execution steps of Figure 5.14 as a *sequence chart*, based on an approach presented in [3]. In general, a chart depicts a sequence of reactions from a single simulation run. Each vertical line in the chart represents a separate species, where time proceeds downward. The vertical line is labeled with the current population of the species, which can evolve

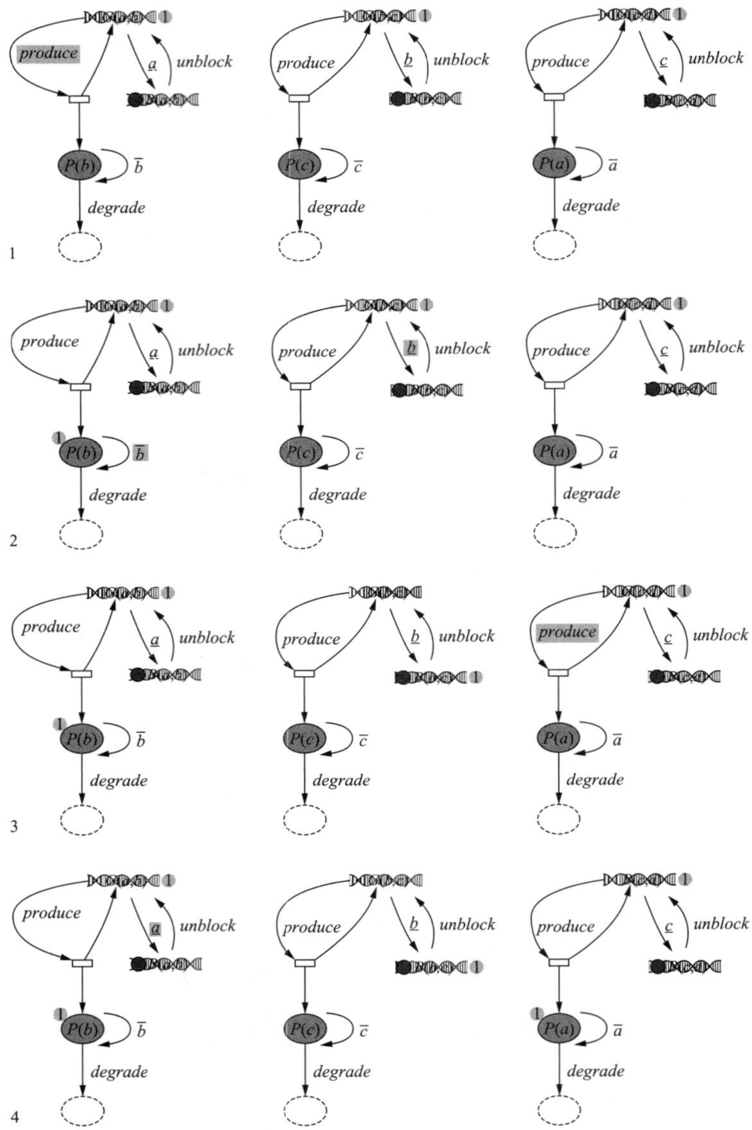

Figure 5.14 A sequence of execution steps for the Repressilator model of Figure 5.12. (1) Initially there is one copy of each gene. The gene $G(a,b)$ produces a protein $P(b)$. (2) Protein $P(b)$ sends on $\overline{b}$ and causes $G(b,c)$ to block. (3) Gene $G(c,a)$ produces a protein $P(a)$. (4) Protein $P(a)$ sends on $\overline{a}$ and causes $G(a,b)$ to block

#	$R(produce)$ $= [G(a,b)]\cdot produce$ $+ [G(b,c)]\cdot produce$ $+ [G(c,a)]\cdot produce$	$R(unblock)$ $= [B(a,b)]\cdot unblock$ $+ [B(b,c)]\cdot unblock$ $+ [B(c,a)]\cdot unblock$	$R(degrade)$ $= [P(b)]\cdot degrade$ $+ [P(c)]\cdot degrade$ $+ [P(a)]\cdot degrade$	$R(a)$	$R(b)$	$R(c)$
1	0.1 + 0.1 + 0.1	0	0	0	0	0
2	0.1 + 0.1 + 0.1	0	0.001 + 0 + 0	0	1·1·1	0
3	0.1 + 0 + 0.1	0 + 0.0001 + 0	0.001 + 0 + 0	0	0	0
4	0.1 + 0 + 0.1	0 + 0.0001 + 0	0.001 + 0 + 0.001	1·1·1	0	0
5	0 + 0 + 0.1	0.0001 + 0.0001 + 0	0.001 + 0 + 0.001	0	0	0
...	0 + 0 + 0.1	0.0001 + 0.0001 + 0	0 + 0 + 100·0.001	0	0	0

Figure 5.15 Propensity calculations for the execution steps of Figure 5.14, where the propensity $\mathcal{R}(a)$ of the a reaction is defined as $[G(a,b)] \cdot [P(a)] \cdot \rho(a)$. The propensities $\mathcal{R}(b)$ and $\mathcal{R}(c)$ are defined in a similar fashion. (1) The simulation starts with one copy of each gene, each of which can produce a protein with propensity 0.1, such that the total propensity of the *produce* reaction is 0.3. (2) A protein $P(b)$ is produced, which can block gene $G(b,c)$ with propensity 1. (3) The gene $G(b,c)$ becomes blocked, preventing the production of protein $P(c)$. But genes $G(c,a)$ and $G(a,b)$ can still each produce with propensity 0.1. (4) A protein $P(a)$ is produced, which can block gene $G(a,b)$ with propensity 1. (5) Now both $G(a,b)$ and $G(b,c)$ are blocked, with a very low propensity to unblock, leaving $G(c,a)$ to produce freely. (...) Eventually an equilibrium is reached between production and degradation of protein $P(a)$, where $\mathcal{R}(produce) = \mathcal{R}(degrade) = 0.1$. This represents the first protein cycle of the simulation

as the simulation progresses. Each horizontal line from one species to another represents a parallel composition of species. The line connects the populations of two or more species, indicating that these populations are all executing simultaneously. Arrows denote reactions that take place as the simulation progresses. Vertical arrows represent delay reactions, while horizontal arrows represent interactions between two species, where the arrow is labeled with the channel on which the interaction takes place. The chart shows how the different genes and proteins interact with one another to initiate the first protein cycle.

5.3.3 3D Visualization

This section describes how the Repressilator model of Figure 5.12 can be simulated in three dimensions, so as to visualize the interactions between genes and proteins over time. The three-dimensional models are displayed using Network 3D.[2] Figure 5.17 shows a 3D representation of a gene with negative control, based on the model in Figure 5.12.

[2]Network 3D was developed by Rich Williams of Microsoft® Research.

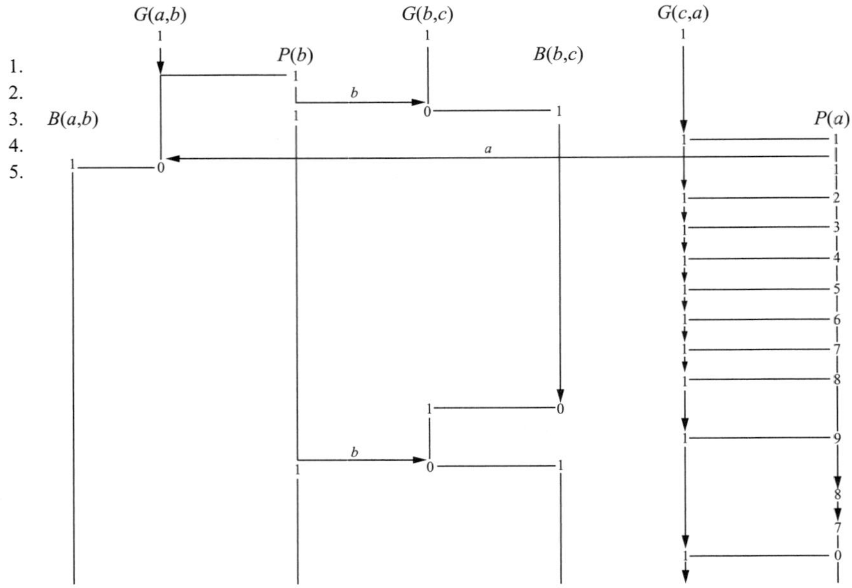

Figure 5.16 A sequence chart of interactions between genes and proteins for the execution steps of Figure 5.14. (1) Initially there is one copy of each of the genes $G(a, b)$, $G(b, c)$, and $G(c, a)$. Gene $G(a, b)$ produces a protein $P(b)$ in parallel with itself. After the production, the populations of $G(a, b)$ and $P(b)$ are both 1. (2) Protein $P(b)$ sends on channel b and interacts with gene $G(b, c)$, which becomes blocked to $B(b, c)$. After the interaction, the populations of $P(b)$, $G(b, c)$, and $B(b, c)$ are 1, 0, and 1, respectively. (3) Gene $G(c, a)$ produces a protein $P(a)$ in parallel with itself. (4) Protein $P(a)$ sends on channel a and interacts with $G(a, b)$, which becomes blocked to $B(a, b)$. (5) Subsequently, $G(c, a)$ produces multiple proteins $P(a)$, some of which are degraded. In addition, $B(b, c)$ unblocks to $G(b, c)$, which is immediately blocked again by $P(b)$

The main difference between the 3D model and the 2D representation given earlier is that each action is now represented as a node instead of as an edge. As a result, the 3D model includes two types of nodes: *action nodes* and *species nodes.* Action nodes are shown in green and are of fixed size, while species nodes are of variable color and size, where the volume of the node is proportional to the population of the species. The node labels are hidden by default, but can be revealed by clicking on individual nodes.

The 3D model also includes two types of edges: *transition edges* and *interaction edges.* Transition edges denote a transition from one species to another, and go from a species node to an action node, and then from an action node to zero or more species nodes. For example, the $\underline{a}$ action has one inbound edge from $G(a, b)$ and one outbound edge to $B(a, b)$, reflecting

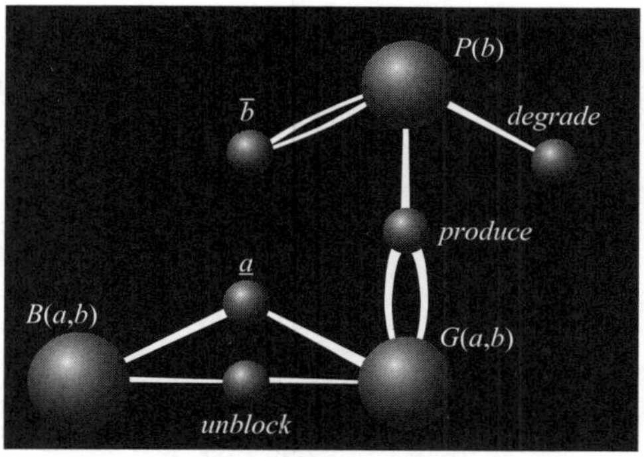

Figure 5.17 A 3D representation of the gene with negative control from Figure 5.12

the fact that the gene can transition from an active state to a blocked state by receiving on $\underline{a}$. The *produce* action has one inbound edge from $G(a, b)$ and two outbound edges to $G(a, b)$ and $P(b)$, as the gene can produce a new protein in parallel with itself. The *degrade* action has only one inbound edge from $P(b)$ and no outbound edges, because the protein disappears after the *degrade* action is executed. Interaction edges denote an interaction between two species, and go from an action node that is sending on a given channel to a corresponding action node that is receiving on the same channel. There are no interaction edges in Figure 5.17, because there are no corresponding send and receive actions on the same channel.

Figure 5.18 shows a 3D representation of the Repressilator model of Figure 5.12, where the definitions of $G(a, b)$, $G(b, c)$, and $G(c, a)$

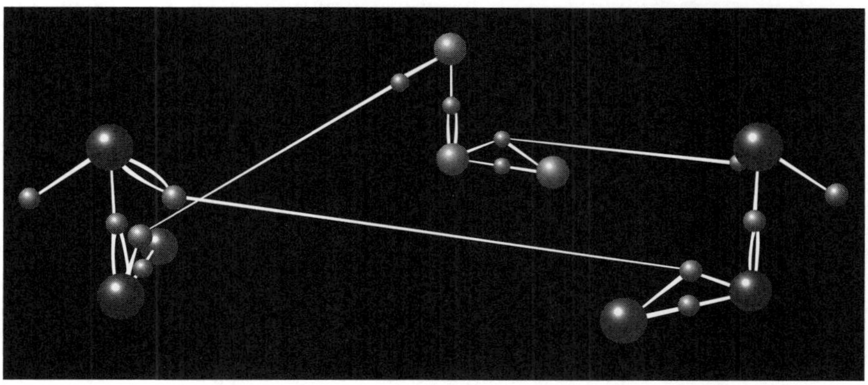

Figure 5.18 A 3D representation of the Repressilator model of Figure 5.12, where the behavior of each gene is represented as a separate subgraph

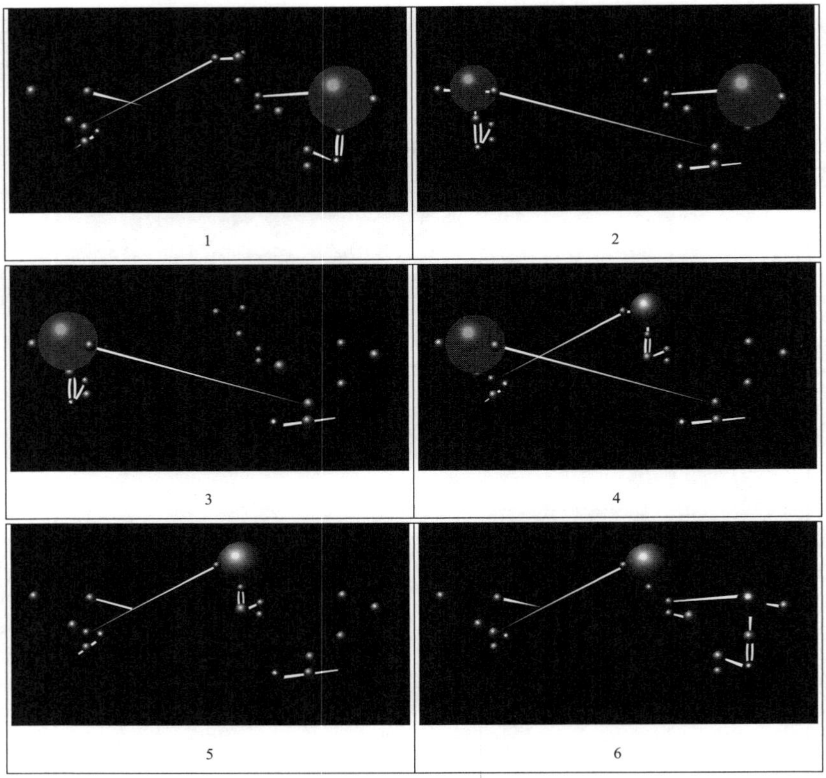

Figure 5.19 A sequence of 3D pictures from a simulation of the Repressilator model of Figure 5.12. (1) Initially there is a large population of red proteins, which switch off the orange gene. (2) The blue gene then switches on and starts producing proteins, which in turn switch off the red gene. (3) Because no more red proteins are produced, the population of red proteins slowly decreases over time, until all of the red proteins are degraded. (4) The orange gene then switches on and starts producing proteins, which in turn switch off the blue gene. (5) Because no more blue proteins are produced, the population of blue proteins slowly decreases over time, until all of the blue proteins are degraded. (6) The red gene then switches on and starts producing proteins, which in turn switch off the orange gene, completing the cycle

are expanded so that the behavior of each gene is represented as a separate subgraph, similar to Figure 5.14. The figure includes three subgraphs, one for each gene, indicated in orange, blue, and red, respectively (see color figures online). The species nodes at the bottom of each subgraph represent active and blocked genes, which have a population of 0 or 1; the species nodes at the top represent proteins, which have variable populations. The subgraphs are connected to one another via interaction edges, which go from the send action of a protein in one subgraph to the receive action of a corresponding gene in another subgraph. For example, an edge goes from the send action $\overline{c}$ of the blue protein to the receive action $\underline{c}$ of the red gene, representing the fact that the blue protein $P(c)$ can switch off the red gene $G(c, a)$.

Figure 5.19 shows a sequence of 3D pictures from a simulation of the Repressilator model of Figure 5.12. The population of a species is visualized by dynamically altering the size of the corresponding species node. When the population is empty, the species node is obscured. The interactions between species are visualized by drawing an edge from a send action to a corresponding receive action on the same channel. When no senders or receivers are present, the interaction edge is obscured. The 3D layout allows complex interactions to be visualized, while avoiding edge crossings that would have been unavoidable in a 2D model. The software can be used to rotate or zoom into the model, so as to focus on a particular subset of interactions and species. The 3D simulation is complementary to the 2D plots of species populations over time, as it allows the interactions between species to be visualized as the simulation progresses.

5.3.4 Model Refinement

Once we understand how the Repressilator model functions, we can further refine the model to improve the regularity of oscillations, as discussed in [2].

We can start by adjusting the rates of the model, as shown in Figure 5.20. We observe that each protein cycle is characterized by a dominant protein whose population stabilizes at an equilibrium between production and degradation, given by $\frac{produce}{degrade}$. In Figure 5.20(a), the average population for a given protein cycle is approximately 100, although this number fluctuates significantly due to stochastic noise. We can limit the relative size of the fluctuations by decreasing the degradation rate to 0.0001, resulting in a dominant population of approximately 1000 proteins, as shown in Figure 5.20(b). We observe that when one protein is dominant, the other two proteins are absent and their corresponding genes are blocked, where one of the blocked genes is actively repressed. For example, when the blue protein is dominant, both the red and orange proteins are absent, where the red gene is actively repressed and the orange gene is waiting to unblock. This situation is illustrated in step 3 of Figure 5.19.

The duration of protein cycles is highly irregular, because it depends mainly on the rate of unblocking the unrepressed gene, which is relatively slow and is characterized by an exponential distribution. We can reduce this variability by increasing the rate of unblocking to $unblock = 0.0001$,

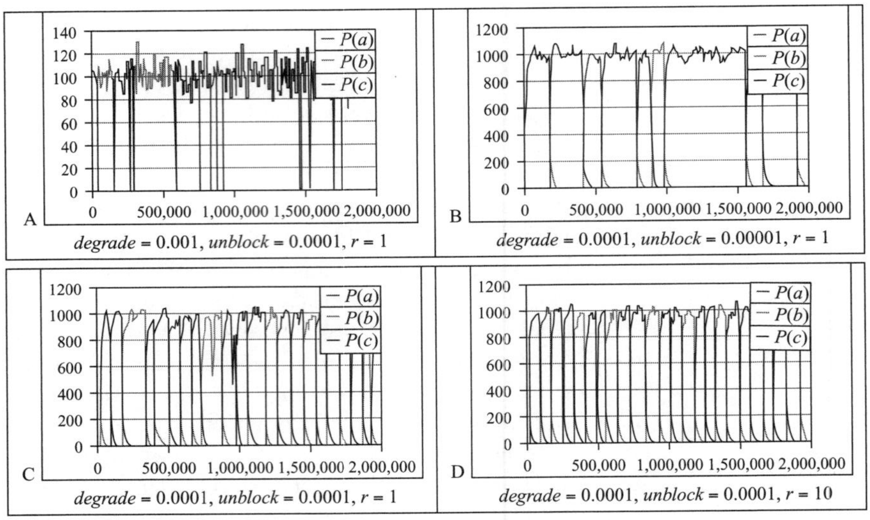

Figure 5.20 Parameter variation for the Repressilator model of Figure 5.12, with *produce* $= 0.1$ and $\rho(a) = \rho(b) = \rho(c) = r$

as shown in Figure 5.20(c). In this setting, the gene unblocks more rapidly, but is immediately blocked again by any remaining repressors. As a consequence of the rapid unblocking, once the last repressor has been degraded, the gene can become active soon thereafter.

Given that the repressor degradation curve is fairly regular, we observe an increased regularity in the oscillations. Unfortunately, because the repressed gene is repeatedly trying to unblock, this also increases the likelihood that it will produce a protein before all of its repressors are degraded, which can result in large fluctuations in protein levels, as observed in Figure 5.20(c). We can compensate for this effect by increasing the rate of gene repression to $r = 10.0$, as shown in Figure 5.20(d). In this setting, even if only a few repressor proteins remain, they will still have a high probability of blocking the corresponding gene, meaning that a gene will become active only after all of its repressors are degraded. This property, together with the regular degradation curve of repressors, gives rise to protein cycles of regular duration and amplitude.

If we were to implement a new version of the Repressilator model inside a bacterium, we might try to use genes and proteins whose reaction rates correspond roughly to those of Figure 5.20(d). The simulation results for such a Repressilator are shown in Figure 5.21, with genes for *tet*, *lambda*, and *lac*, together with a reporter gene for *gfp*. Note that GFP is produced in absence of the *tet* protein, where the intermittent expression of GFP corresponds to the bacterium repeatedly glowing on and off.

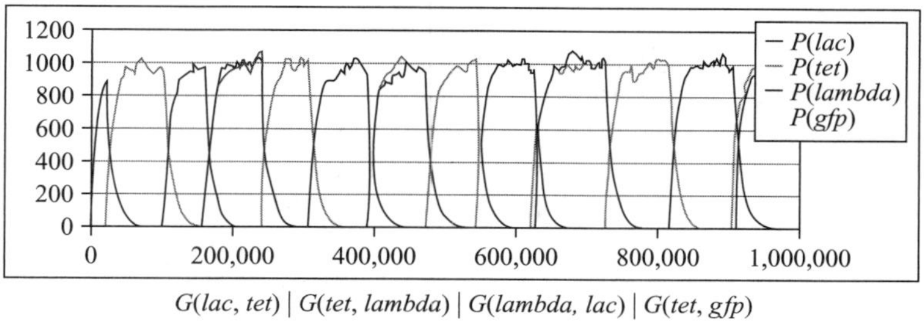

Figure 5.21 Simulation results for the Repressilator model of Figure 5.12, with an additional GFP reporter, where *produce* $= 0.1$, *degrade* $=$ *unblock* $=$ 0.0001, and $r = 10$

In addition to adjusting the rates of the model in Figure 5.12(a), we can refine the model to include transcription and translation steps explicitly, together with repressor binding and unbinding, as shown in Figure 5.22(a). The gene $G(a, b)$ can *transcribe*, after which a new mRNA strand $M(b)$ is

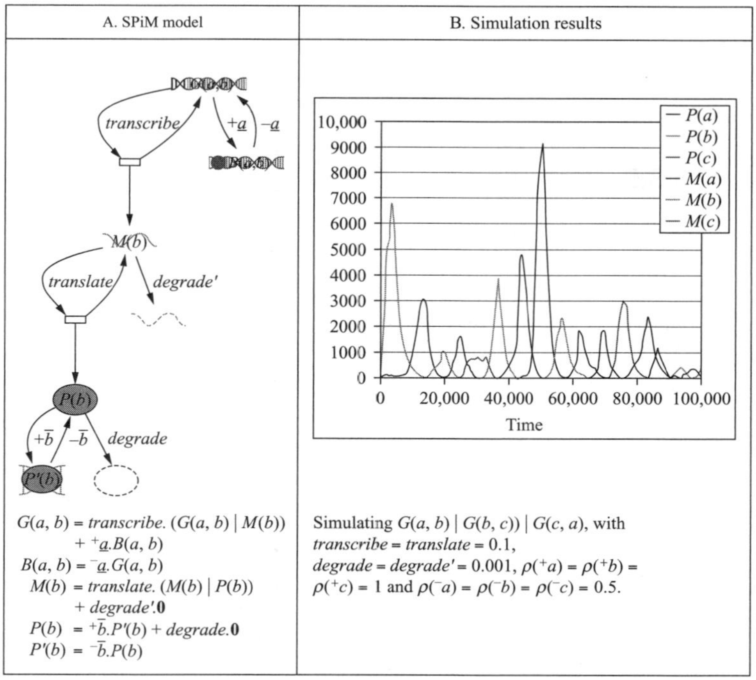

Figure 5.22 Refining the gene with negative control of Figure 5.12 to include transcription, translation, and repressor binding

produced in parallel. Alternatively, the gene can receive and bind on $^{+}\underline{a}$ and become blocked by a repressor, after which it can receive and unbind on $^{-}\underline{a}$ and release the repressor, becoming active once more. The mRNA strand $M(b)$ can *translate* and produce a new protein $P(b)$ in parallel. Alternatively, the strand can *degrade*. The protein $P(b)$ can send and bind on $^{+}\overline{b}$ and become bound to a gene, after which it can send and unbind on $^{-}\overline{b}$. Alternatively, the protein can *degrade*. Simulation results for this model are shown in Figure 5.22(b). Note that the protein levels fluctuate much more widely than in Figure 5.12(b), as they amplify any small fluctuations that occur in the levels of mRNA.

Although the detailed model in Figure 5.22 is much more complex than the simplified model of Figure 5.12, similar analysis can still be applied to understand how the model functions. However, further refinements to the model are required to reduce the fluctuations in mRNA levels. One way of reducing fluctuations is to introduce some form of cooperativity in the model, by allowing proteins to form dimers or even tetramers before repressing a gene. We omit the extensions here, but additional details can be found in [2]. The main point is that we can continually refine our parameterized model of a single gene without changing the structure of the overall gene network, which remains fixed at $G(a, b) \mid G(b, c) \mid G(c, a)$. This example illustrates the modularity of our approach, where module definitions can be continually refined without changing the main program.

5.3.5 Genetic Logic Gates

A substantial extension to the original Repressilator experiment is described in [14], in which a collection of logic gates were implemented in live bacteria. Rather than producing oscillations in GFP levels, this experiment aimed to construct gene networks that were able to function as Boolean logic gates. The extension was achieved by introducing inducers aTc and IPTG to inhibit the *tet* and *lac* proteins, respectively. The presence or absence of the inducers represented the logical inputs to the network, while the presence or absence of GFP represented its logical output. In total, 125 different networks of three promoter-gene units were constructed, using a combination of three different genes and five different promoters. For each of these networks, the level of GFP was measured in the presence or absence of aTc, IPTG, or both.

We can model these networks in the SPiM calculus by defining a parameterized gene in which the produced proteins are switched off by an inducer, as shown in Figure 5.23(a). The process $Gi(a, b, i)$ represents a gene that is blocked by receiving on $\underline{a}$, and produces proteins that send on $\overline{b}$, where the produced proteins are inhibited by receiving on $\underline{i}$. We can use the gene $Gi(a, b, i)$ together with the gene $G(a, b)$ of Figure 5.12 to construct

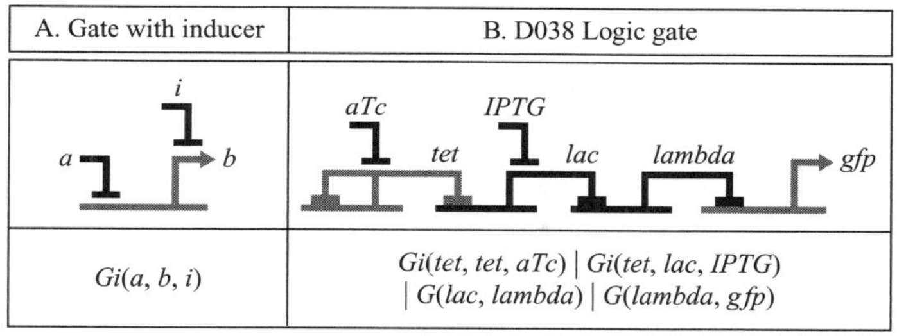

Figure 5.23 Modular construction of gene networks from two parameterized models of a gene. (a) A parameterized gene whose proteins are inhibited by an inducer i. (b) A gene network that behaves like a Boolean logic gate. The proteins aTc and IPTG act as Boolean inputs, while GFP acts as a Boolean output. Depending on the presence or absence of inhibitors, the GFP levels are either high or low, respectively. When this model is implemented in live bacteria, the presence or absence of inducers determines whether the bacteria will glow

the networks described in [14]. One of these networks is illustrated in Figure 5.23(b). The first gene in the network is inhibited by *tet* and produces *tet*, the second gene is inhibited by *tet* and produces *lac*, and the third gene is inhibited by *lac* and produces *lambda*. Finally, the network is fixed so that the *lambda* gene inhibits *gfp*, meaning that the bacteria glows in the absence of *lambda*. The two inputs to the system are aTc and IPTG, where aTc inhibits the *tet* proteins and IPTG inhibits the *lac* proteins.

Figure 5.24(a) shows the definition of the parameterized gene with inducers. The definition is similar to that of Figure 5.12, except that the gene is parameterized by an inducer i, which inhibits the produced proteins. The protein $Pi(b, i)$ can block a gene by sending on $\overline{b}$, or it can be inhibited by receiving on $\underline{i}$. Alternatively, the protein can *degrade*. The behavior of an inducer $I(i)$ is straightforward, in that it can inhibit a protein by sending on $\overline{i}$.

Figure 5.24(b) shows the simulation results for the gene network of Figure 5.23(b), where the green line represents the levels of GFP. Each plot represents a simulation of the network with different combinations of inputs. We observe that GFP is high in the presence of $I(aTc)$ and in the absence of $I(IPTG)$, which gives a logical characterization of the network. The outcome of the simulation is not immediately obvious by looking at the network construction of Figure 5.23, not least because of the presence of a *head feedback* loop, where one of the genes is repressed by the protein it produces. A more detailed explanation of the simulation results is given in [1]. The main point to note here is that we can model a wide range of genetic logic gates using a small number of simple genetic building blocks.

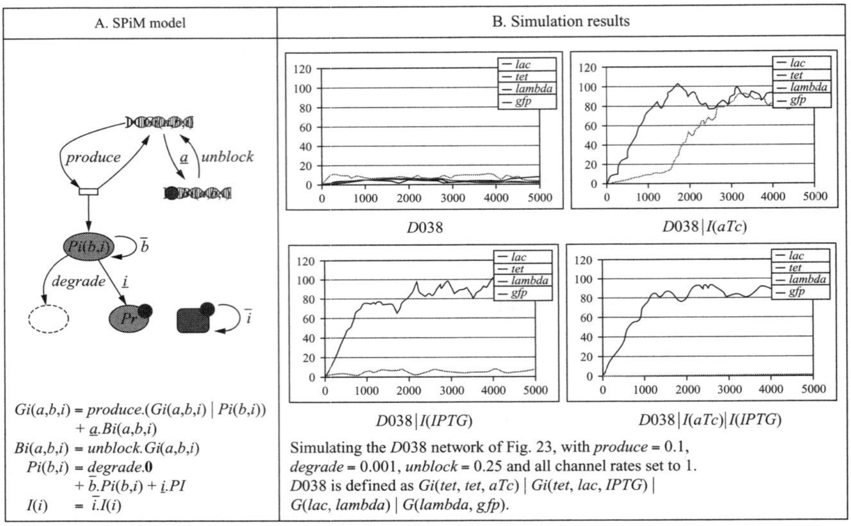

Figure 5.24 A SPiM model of a parameterized gene with negative control and protein inhibition (a), with associated simulation results (b). This model is similar to that of Figure 5.12, except that here the gene is parameterized by an inducer i, which inhibits the produced proteins

5.4 *C. elegans* Development

The biological examples presented so far have all been modeled at the level of interacting genes and proteins. In this section, we explore how the SPiM calculus can be used to model systems at the level of interacting cells. We consider an example from developmental biology, relating to the development of the egg-laying system of the *C. elegans* nematode. *C. elegans* has been used to study a number of fundamental cellular processes, and many breakthroughs relating to human physiology were first discovered in this organism. Examples include genetic regulation of organ development and programmed cell death, as well as RNA interference–gene silencing by double-stranded RNA; the investigations of these two lines of research were awarded the Nobel Prize in Physiology or Medicine in 2002 and 2006, respectively. More recently, the Nobel Prize in Chemistry was awarded for the discovery and development of the green fluorescent protein, GFP, for which a significant part of the research was carried out in *C. elegans*.

In this section we consider the phase of development in *C. elegans* in which six vulval precursor cells (VPCs) decide to adopt a primary, secondary, or tertiary fate depending on their positions with respect to an anchor cell (AC). The decision of a precursor cell to adopt a particular fate

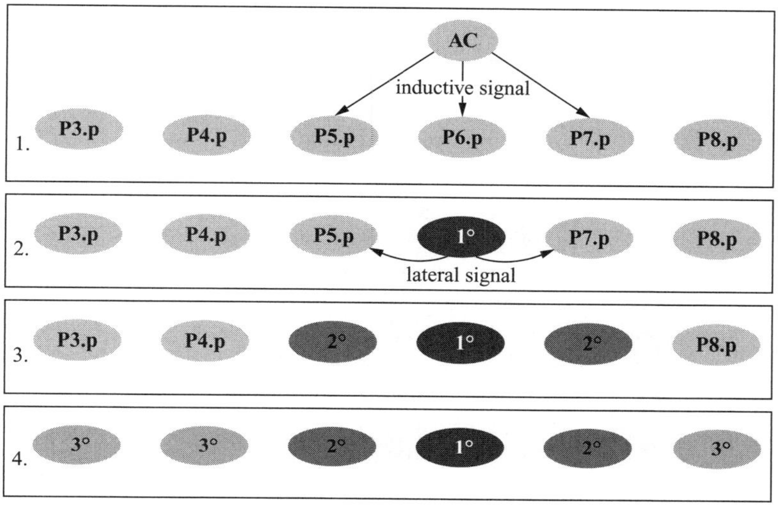

Figure 5.25 Some of the main events that take place during normal differentiation of VPC cells in *C. elegans.* (1) Initially the six cells P3.p–P8.p are undifferentiated. The anchor cell sends an inductive signal, which is strongest for the cell P6.p directly in front of it. (2) The strong inductive signal causes cell P6.p to adopt a primary fate, after which it sends a lateral signal to its immediate neighbors. (3) The lateral signal causes cells P5.p and P7.p to adopt a secondary fate. (4) Eventually, the remaining cells adopt a tertiary fate

is an important part of the development of the organism, which will ultimately determine whether the adult worm can successfully lay eggs. More details about the system can be found in [12, 36, 37] and a recent review is presented in [35]. Various models of VPC differentiation have been published recently, such as [9, 12, 21] to cite a few, in which the differentiation process has been highly simplified. Here we present an even simpler model, which illustrates some of the main factors involved during development. While many biological details are missing from this example, our approach allows further model refinements to be made so that relevant details can be included incrementally.

Figure 5.25 depicts some of the main events that take place during normal differentiation of VPC cells in *C. elegans.* Once the cells have adopted their respective fates, they undergo further stages of proliferation and differentiation to develop into the egg-laying system of the adult worm.

5.4.1 Individual Cell Models

We start by constructing a simplified model of an anchor cell, together with a generic model of an undifferentiated precursor cell that can adopt one of three fates: primary, secondary, or tertiary. The cell models are given in Figure 5.26, where the graphical representation at the top is equivalent to the textual representation at the bottom. The anchor cell AC can repeatedly send on $\overline{ac}$, which represents a persistent inductive signal. A precursor cell can have four possible states, representing the undifferentiated cell V, the primary cell fate $V1$, and the secondary and tertiary cell fates $V2$ and $V3$, respectively.

The undifferentiated cell V is parameterized by its position x inside the worm, which is represented by a number between 3 and 8. Cells with successive numbers are considered to be adjacent to each other, and the cell at position 6 is considered to be directly opposite the anchor cell. The undifferentiated cell is also parameterized by a rate k, which represents the strength of its interaction with the anchor cell. Cells with a low k are considered far away and interact only weakly with the anchor cell, while cells with a high k are considered nearby and interact strongly with the anchor cell. For simplicity, we assume only three levels for k: *low*, *medium*, and *high*. Each undifferentiated cell is also characterized by its left and right neighbors, given by the channels l and r, respectively, which correspond to the lateral signals that a cell can use to communicate with its neighbors. The intuition is that each cell will interact with its left neighbor on channel l and with its right neighbor on channel r. We assume that once the precursor cells have adopted a particular fate, they no longer interact with the anchor cell.

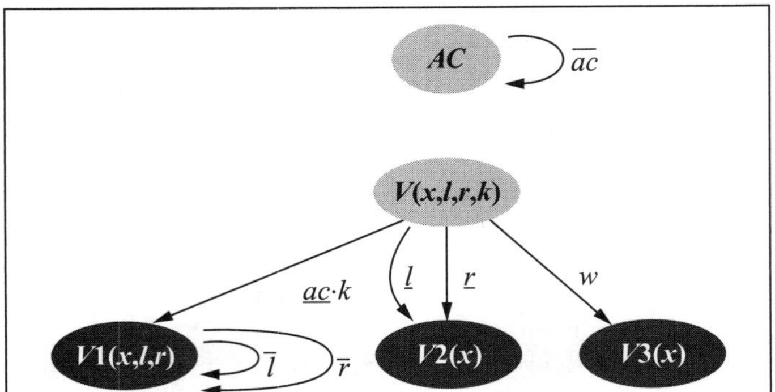

$$AC = \overline{ac}.\,AC$$
$$V(x, l, r, k) = \underline{ac}{\cdot}k.\ V1(x, l, r) + l.V2(x) + \underline{r}.V2(x) + w.V3(x)$$
$$V1(x, l, r) = \overline{l}.V1(x, l, r) + \overline{r}.V1(x, l, r)$$

Figure 5.26 A SPiM model of an anchor cell and a vulval precursor cell, which are instrumental in the development of the egg-laying system of the *C. elegans* nematode. The undifferentiated precursor cell V can adopt one of three fates: a primary fate $V1$, a secondary fate $V2$, or a tertiary fate $V3$

Initially, a precursor cell starts out undifferentiated as V. One of three things can then happen:

1. It can receive a signal on $\underline{ac}$ at rate k from the anchor cell and adopt a primary fate $V1$. The primary cell will then send a persistent signal to its left and right neighbors on $\overline{l}$ and $\overline{r}$, respectively.
2. It can receive a signal from its left or right neighbor on $\underline{l}$ or $\underline{r}$, respectively, and adopt a secondary fate $V2$.
3. In the absence of a stimulus, it can eventually adopt a tertiary fate $V3$, at rate w.

5.4.2 System Model

A model of the full developmental system is shown in Figure 5.27, consisting of the anchor cell and the six VPC cells. In the figure, the graphical representation at the top is equivalent to the textual representation at the bottom. Each of the VPC cells is an instance of the generic model of a single cell described in Figure 5.26. The generic model is instantiated with different parameters depending on the position of the cell, the strength of its interaction with the anchor cell, and the neighboring cells with which it can communicate. Cells are assumed to communicate with their left and right neighbors on dedicated channels, where the left channel of a given cell is equal to the right channel of its left neighbor. For example, the cell $V(6, s6, s7, high)$ at position 6 can communicate with its left neighbor on channel $s6$ and with its right neighbor on channel $s7$. It can also communicate with the anchor cell at a *high* rate, because it is assumed to be in close proximity to the anchor cell. Note that channels $s3$ and $s9$ in

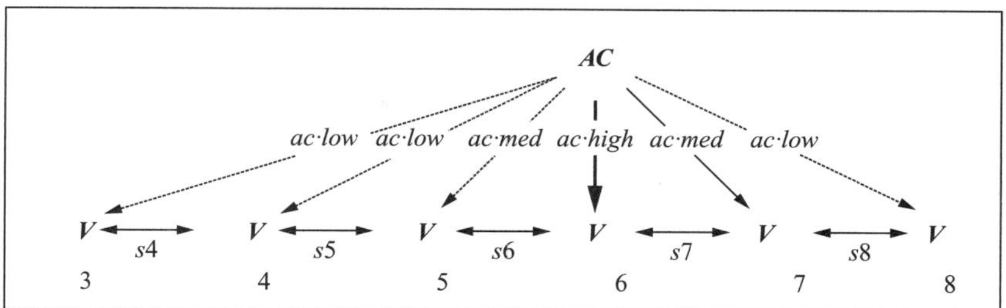

AC | *V*(3, *s*3, *s*4, *low*) | *V*(4, *s*4, *s*5, *low*) | *V*(5, *s*5, *s*6, *med*)
| *V*(6, *s*6, *s*7, *high*) | *V*(7, *s*7, *s*8, *med*) | *V*(8, *s*8, *s*9, *low*)

Figure 5.27 System model of cell–cell interactions during VPC differentiation in *C. elegans*. The model is obtained by placing an anchor cell in parallel with six different VPC cells, using the generic model of a VPC cell described in Figure 5.26

Figure 5.27 are unused, since the cells to the left of position 3 and to the right of position 8 are not represented. The model can be further extended to represent these cells if necessary.

5.4.3 Simulation Results

To simulate the process of VPC differentiation, we assign approximate rates to the generic VPC model of Figure 5.26 and to the overall system in Figure 5.27. The rates of *low*, *medium*, and *high* inductive signaling are given by 0.0, 1.0, and 10.0, respectively, while the rate ls of lateral signaling is given by 1.0. We assume that the rate of lateral signaling is the same for all precursor cells, with the rates of channels $s4, \ldots, s8$ being equal to ls. The rate w for a cell to adopt a tertiary fate in the absence of external signals is given by 0.0001. Although these rates are entirely arbitrary, they provide an initial approximation for the relative speeds of the various interactions that take place during development, and they can be adjusted accordingly as new insight is gained from simulation and experimentation.

The simulation results of Figure 5.28 show the total population of undifferentiated cells V and of differentiated cells $V1, V2$, and $V3$ over time. Initially there are six undifferentiated cells. One of these cells adopts a primary fate almost immediately. Two of the cells then adopt secondary fates in quick succession, and the remaining cells eventually adopt a tertiary fate. To determine which cells adopt which fates, we plot the number of cells $V, V1, V2$, and $V3$ at each of the six positions as shown in Figure 5.29, where each circle represents a precursor cell. The horizontal coordinate denotes the cell position, which ranges from 3 to 8. The color denotes the cell fate, where an undifferentiated cell V is shown in gray, and a cell that adopts a primary, secondary, or tertiary fate is shown in red, blue, or yellow, respectively. The plots are generated automatically during the simulation, by using the parameter x in processes $V, V1, V2$, and $V3$ as the horizontal

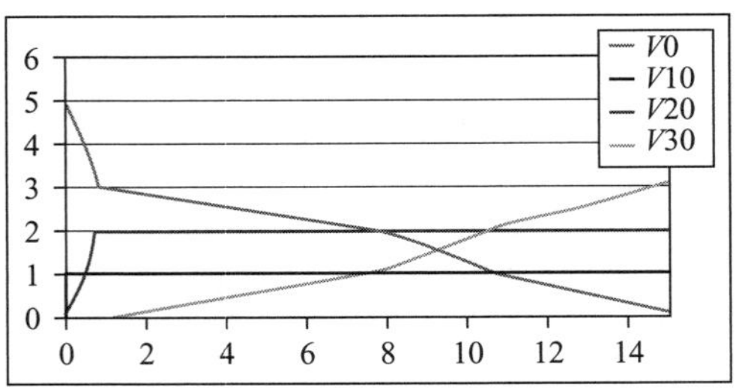

Figure 5.28 Simulation results for the VPC model of Figure 5.27. The results show the populations of undifferentiated, primary, secondary, and tertiary cells over time.

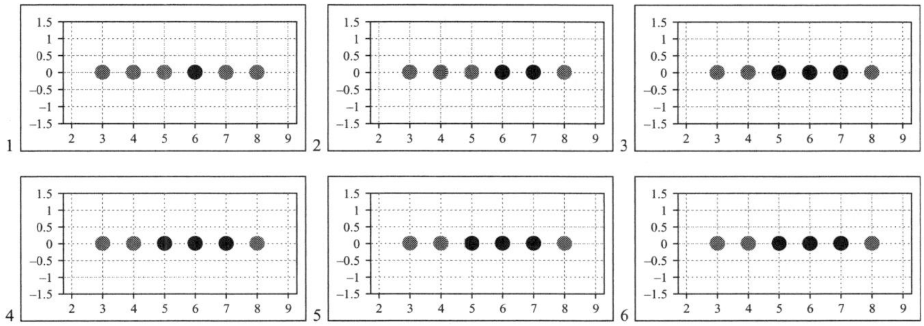

Figure 5.29 Coordinate geometry plots for the simulation results of Figure 5.28. The results show the change in cell fate over time for cells at positions 3 to 8, where the horizontal coordinate denotes the cell position. Initially, all cells start out as undifferentiated (not shown). The cell at position 6 then adopts a primary fate, after which the cells at positions 7 and 5 adopt a secondary fate. One by one, the remaining cells then adopt a tertiary fate

coordinate for each process. As a given cell adopts a particular fate, the process $V(x,l,r,k)$ of each cell transitions to either $V1(x,l,r)$, $V2(x)$, or $V3(x)$, and the processes at each position x are plotted accordingly. In this way, the SPiM simulation generates a movie of cell differentiation over time. We can take snapshots of the simulation to separate out the individual steps, as shown in Figure 5.29. For example, the first frame plots processes $V(3,s3,s4,low) \mid V(4,s4,s5,low) \mid V(5,s5,s6,med) \mid V1(6,s6,s7) \mid V(7,s7,s8,med) \mid V(8,s8,s9,low)$, with process $V1(6,s6,s7)$ at position 6 representing a primary fate and the remaining processes representing undetermined fates. As the simulation progresses, the processes interact with one another according to the behavior outlined in Figure 5.26, with the processes $V(7,s7,s8,med)$ and $V(5,s5,s6,med)$ transitioning to $V2(7)$ and $V2(5)$, respectively, followed by the remaining processes transitioning to $V3(4)$, $V3(8)$, and $V3(3)$.

We can interpret the simulation results based on the model construction and the chosen parameters. The precursor cell at position 6 adopts a primary fate by interacting with the anchor cell at rate $high = 10$. The average duration of this interaction is 0.1, which explains the rapid transition time observed in the simulation. Next, the primary cell at position 6 signals to its left and right neighbors to adopt a secondary fate at rate $ls = 1.0$. The average duration of these transitions is 1, which is 10 times slower than the initial inductive signal to the cell at position 6. In the absence of external signals, the remaining cells adopt a tertiary fate at rate $w = 0.0001$, which explains the length of time taken for these cells to differentiate.

Although the rates were chosen arbitrarily in this example, if we increase the rate w, then the cells will be able to adopt a tertiary fate more quickly. Because the simulation is stochastic, this rate change means that the cells at positions 5 to 7 will have a greater chance of adopting a tertiary fate, even in the presence of inductive and lateral signals, which could potentially result in abnormal development. Note also that the cells at positions 5 and 7 receive an inductive signal at rate 1, which is 10 times weaker than the inductive signal to the cell at position 6. Although the inductive signal is much weaker, there is still a 1 in 10 chance that a cell at position 5 or 7 might adopt a primary fate, resulting in abnormal development. Indeed, if we simulate the model several times, we observe that cells 5 and 7 adopt the wrong fate roughly 10 percent of the time. Clearly, the VPC model is not satisfactory, because in reality the process of VPC differentiation is extremely robust. Simply changing the ratio of the $high$ to med inductive signals to something like 1000 to 1 in an attempt to improve robustness is not acceptable either, because in reality the strength of the inductive signal does not differ by such a large amount between neighboring cells. Thus, although the high-level model we have presented can mechanistically reproduce some of the stages of VPC differentiation, it cannot reproduce the inherent robustness of the process, and more refined models are needed.

5.4.4 Refined Model

To illustrate our modeling approach, we describe a refined version of the VPC model of Figure 5.26, based on the early results of [37]. The refined model is presented in Figure 5.30. One of the simplifying assumptions of the initial model was to represent a change in cell fate by the activation of a single process inside the cell—representative, for example, of the activation of a single gene. In the refined model, we now represent a change in cell fate by the accumulation of a particular protein, where $V1, V2$, and $V3$ represent proteins with primary, secondary, and tertiary functions, respectively. As with the simplified model, an undifferentiated cell V is parameterized by its position x, the strength k of its interaction with the anchor cell, and the channels l and r on which it can communicate with its left and right neighbors, respectively. As a result, the overall model of the VPC system remains the same as in Figure 5.27, while the behavior of an undifferentiated cell $V(x, l, r, k)$ is substantially refined. Instead of consisting of a single process that changes state over time, the cell V now consists of multiple processes running in parallel. In Figure 5.30, the processes $gV1, gV2$, and $gV3$ represent the genes for proteins $V1$, $V2$, and $V3$, respectively, while the processes $gVul$ and $gLin12$ represent the genes for proteins Vul and $Lin12$, respectively. The genes are executed in parallel with each other inside the cell, alongside the Muv proteins.

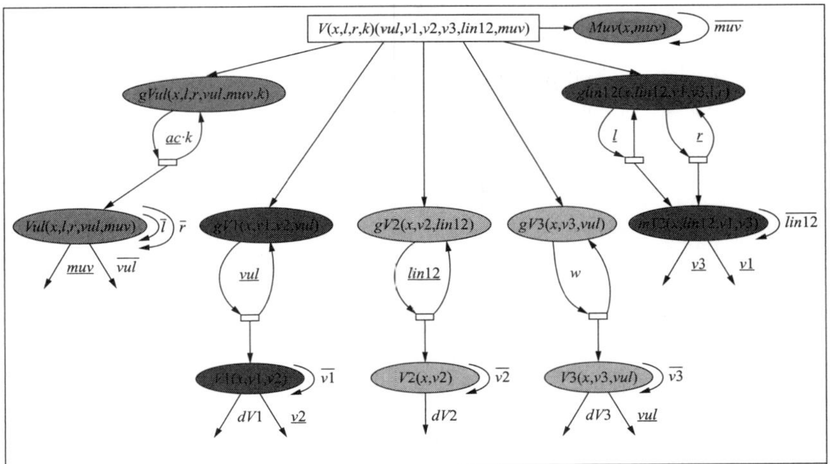

$$
\begin{aligned}
V(x, l, r, k) &= (vul, v1, v3, v3, lin12, muv)\\
&\quad (gVul(x, l, r, vul, muv, k) \mid gLin12(x, lin12, v1, v3, l, r)\\
&\quad \mid gV1(x, v1, v2, vul) \mid gV2(x, v2, lin12) \mid gV3(x, v3, vul) \mid Muv(x, muv))\\
gVul(x, l, r, vul, muv, k) &= \underline{ac}{\cdot}k{\cdot}(gVul(x, l, r, vul, muv, k) \mid Vul(x, l, r, vul, muv))\\
gLin12(x, lin12, v1, v3, l, r) &= \underline{l}.(gLin12(x, lin12, v1, v3, l, r) \mid Lin12(x, lin12, v1, v3)\\
&\quad + \underline{r}.(gLin12(x, lin12, v1, v3, l, r) \mid Lin12(x, lin12, v1, v3)\\
gV1(x, v1, v2, vul) &= \underline{vul}.(gV1(x, v1, v2, vul) \mid V1(x, v1, v2))\\
gV2(x, v2, lin12) &= \underline{lin12}.(gV2(x, v2, lin12) \mid V2(x, v2))\\
gV3(x, v3, vul) &= \underline{w}.(gV3(x, v3, vul) \mid V3(x, v3, vul))\\
Vul(x, l, r, vul, muv) &= \overline{l}.Vul(x, l, r, vul, muv) + r.Vul(x, l, r, vul, muv) + \overline{vul} + muv\\
Lin12(x, lin12, v1, v3) &= \overline{lin12}.Lin12(x, lin12, v1, v3) + \underline{v3} + \underline{v1}\\
Muv(x, muv) &= \overline{muv}.Muv(x, muv)\\
V1(x, v1, v2) &= \overline{v1}.V1(x, v1, v2) + \underline{v2} + dV1\\
V2(x, v2) &= \overline{v2}.V2(x, v2) + dV2\\
V3(x, v3, vul) &= \overline{v3}.V3(x, v3, vul) + \underline{vul} + dV3
\end{aligned}
$$

Figure 5.30 A refined SPiM model of the vulval precursor cell of Figure 5.26. The proteins $V1, V2$, and $V3$ represent primary, secondary, and tertiary functions, respectively, that correspond to particular cell fates

The gene $gVul$ produces proteins in the presence of a persistent inductive signal on channel ac, where the rate of production is determined by the parameter k. The Vul proteins then initiate a persistent communication with neighboring cells by sending on $\overline{l}$ and $\overline{r}$. The Vul proteins can also induce the production of proteins $V1$ by sending on $\overline{vul}$. The proteins $V1$ are specific to cells that adopt a primary fate and can persistently send on $\overline{v1}$, which causes the $Lin12$ receptors to degrade. This prevents the cell from responding to its own lateral signals on $\underline{l}$ and $\underline{r}$, thereby preventing it from adopting a secondary fate. Conversely, the gene $gLin12$ produces active receptor proteins when it receives a signal on $\underline{l}$ or $\underline{r}$ from its left or right neighbors, respectively. The $Lin12$ receptors send on $\overline{lin12}$ and cause the production of proteins $V2$, which are specific to cells that adopt a secondary fate. The proteins $V2$ also cause the degradation of proteins $V1$ by sending on $\overline{v2}$, thereby preventing the cell from adopting a primary fate. Finally, the gene $gV3$ is constitutively expressed at rate w and produces proteins $V3$, which are specific to cells with a tertiary fate. These proteins are, in turn, inhibited by receiving on $\underline{vul}$.

Thus there is a complicated network of interactions to ensure that the cells reliably adopt specific fates depending on the signals they receive. The interactions inside a particular cell take place on channels $vul, v1, v2, v3, lin12$, and muv. To ensure that the signals inside a given cell do not interfere with those inside neighboring cells, the channels are *restricted* to each cell, denoted in the figure by placing them in brackets inside the cell definition. It is this notion of restricted channels that enables boundaries between cells to be established in the model.

Initial simulation results for the VPC model of Figure 5.30 are presented in Figure 5.31. The results show the populations of processes $V1$, $V2$, and $V3$ for each of the six cells. Because these processes are parameterized by their location x, there are 18 different processes altogether. The first three lanes represent populations of $V1, V2$, and $V3$ for the cell at position 3, the next three lanes represent populations of proteins for the cell at position 4, and so forth. We observe that the cell at position 6 adopts a primary fate, as it has a predominance of proteins $V1$, while the cells at positions 5 and 7 adopt a secondary fate, with a predominance of $V2$ proteins. Finally, the

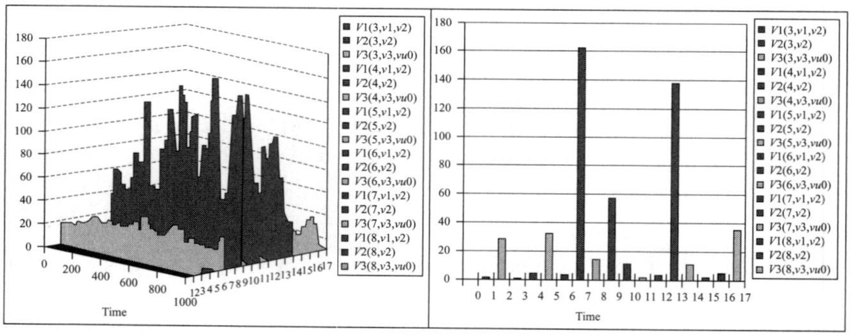

Figure 5.31 Simulation results for the VPC model of Figure 5.30, with $low = 0.01$, $med = 2$, $high = 10$, $v1 = 10$, $v2 = 0.1$, $v3 = 1$, $dV1 = 0.1$, $dV2 = dV3 = 0.05$, and all remaining rates set to 1. The results indicate the populations of processes $V1$, $V2$, and $V3$, which represent primary, secondary, and tertiary fates, respectively, for each of the six cells. Both plots represent different views of the same simulation. The left plot represents a 3D view of the 18 different protein concentrations over time, three for each cell. The right plot gives another view of the same results, which displays the maximum populations for each of the proteins. Each cell has a dominant population of a particular protein, which corresponds to a particular cell fate

cells at positions 3, 4, and 8 adopt a tertiary fate, with a predominance of $V3$ proteins. Rather than making a decision after receiving a single signal from a neighboring cell, repeated signaling is required for a cell to adopt a particular fate. This improves the robustness of the system, as the decision to adopt a particular fate is now made over a period of time. An important mechanism in this process is the amplification of a relatively small difference in inductive signal between cells, so that each cell can reliably adopt the correct fate. Some of these issues are discussed in more detail in [9].

The main point made here is that we can start with a highly simplified model of a cell, then instantiate multiple copies of this model with different parameters to represent a system of communicating cells. We can then refine the model of a cell without changing the overall system definition. Additional details can be included in subsequent refinements, based on published results [9, 12, 21, 35]. This highlights the main motivation for our approach, which is to construct complex systems from simpler building blocks, and to be able to refine these building blocks without having to reconstruct the entire system.

5.5 SPiM with Compartments

The previous section described how the SPiM calculus can be used to model interactions between cells with fixed locations. In this section we describe how the calculus can be extended with a notion of *mobile compartments* so as to model interactions between cells that can move relative to each other. The ambient calculus by Cardelli and Gordon [5] originally introduced a notion of mobile compartments to model computation carried out in mobile devices, together with mobile code that can move between devices. Since the original calculus was outlined, numerous variants have been introduced to model various aspects of security and mobility of distributed computer systems. More recently, the bioambient calculus was introduced by Regev and colleagues [31] to model molecular localization and compartmentalization in biological systems. In this section we present an extension of the SPiM calculus with compartments, which is essentially a graphical variant of the bioambient calculus.

5.5.1 Basic Primitives

We introduce the basic primitives of the SPiM calculus with compartments by means of a simple example. The example models some of the interactions that can take place between cells of the human immune system; it represents only a tiny fraction of the processes that actually take place during an immune response.

Dendritic cells are part of the *adaptive immune system*, one of the main defense mechanisms of the human body against viral infections. Dendritic cells act as sentinels, patrolling areas of the body that can be exposed to infection, such as the lungs and the tissue underneath the skin or surrounding the intestines. If a dendritic cell becomes infected by a virus such as measles, the virus hijacks some of the internal machinery of the cell to produce more copies of itself, consuming some of the resources of the cell in the process. When it is infected with a virus, a dendritic cell can detect the infection by means of internal receptors, which cause the cell to become activated.

Once activated, the dendritic cell can travel to a nearby *lymph node*, where it presents fragments of some of the proteins being made by the virus. These fragments are presented by means of *MHC class I molecules*, which capture the fragments from inside the cell and present them at the cell surface. The MHC class I molecules provide an external signal that the cell is infected, by presenting specific parts of the viral proteins as evidence of the infection.

Once inside a lymph node, the dendritic cell with MHC class I molecules on its surface will encounter a large number of naive *CD8 T cells*, some of which will recognize the viral protein fragments being presented and become activated as a result. Only a very small proportion of the total population of CD8 T cells will be able to recognize a particular protein fragment from a particular virus, however, because the T-cell repertoire needs to be broad enough to recognize potentially billions of viral proteins. This broad repertoire is achieved by a process of *genetic recombination*, involving a random mix-and-match of different genetic sequences during T-cell development. Thus each T cell develops receptors that can recognize only certain protein fragments.

When the receptors of a given CD8 T cell recognize a particular protein fragment being presented on the surface of an infected dendritic cell, the T cell can become activated, provided it also receives the necessary co-stimulatory signals. The activated CD8 T cell then proliferates to make more copies of itself, creating an army of such cells to effectively handle the viral infection. The activated T cells leave the lymph node, traveling through the bloodstream until they reach the site of infection. During their journey, they are guided by specific *chemokine*-signaling molecules, while specific *addressin* and *selectin* molecules allow them to identify and enter the site of infection.

Upon their arrival, the activated CD8 T cells are able to use their receptors to target the infected cells. This targeting is made possible by the fact that the infected cells are also presenting protein fragments from the virus with which they are infected, by means of MHC class I molecules. The T-cell army can recognize the specific protein fragments being presented by the infected cells, and can target those cells for destruction to help contain the viral infection.

The preceding description includes only a tiny fraction of the events that actually occur during a CD8 T-cell response. More details are available in standard textbooks [20, 34], some of which are searchable online.[3] A recent review of the presentation of protein fragments by MHC class I molecules can also by found in [38].

Figure 5.32 presents a SPiM model of some of the basic processes that can take place during an immune response to a viral infection. The graphical representation at the top is equivalent to the textual representation at the bottom. The example illustrates the main primitives of the SPiM calculus with compartments. We model the site of infection as a particular compartment, which represents a region of tissue cells belonging to an infected organ (such as the liver or a lung). The compartment is drawn as a box around a collection of processes, all of which are executing in parallel inside the compartment. We also model a cell as a compartment, with its own internal processes. In this case, the compartment is a means of separating the computation taking place inside the cell boundary from the computation taking place outside the cell.

When a dendritic cell becomes infected, it can become activated and migrate from the site of infection toward a *secondary lymphoid organ*, such as a nearby lymph node. In the SPiM calculus, this movement is represented by the `out` $\overline{tissue}$ action. The `out` keyword denotes the direction of movement, while the send on $\overline{tissue}$ denotes the channel over which the movement takes place. In parallel, the tissue contains a persistent `out` $\underline{tissue}$ action, which allows cells to leave the tissue. Thus the send `out` $\overline{tissue}$ is executed inside the compartment that is leaving, while the receive `out` $\underline{tissue}$ is executed inside the compartment that is allowing the leave to take place. The sender and receiver can then interact on channel *tissue*, after which the sender compartment moves outside the receiver compartment.

Note that the entire compartment moves together with all of its contents after the interaction takes place. The dendritic cell is then in a *Migrating* state and can enter a lymph node by the `in` $\overline{lymph}$ action. The `in` keyword denotes the direction of movement, while the send on $\overline{lymph}$ denotes the channel over which the movement takes place. In parallel, the lymph node contains a persistent `in` $\underline{lymph}$ action, which allows cells to enter the lymph node. The send `in` $\overline{lymph}$ is executed by the compartment that is entering, while the receive `in` $\underline{lymph}$ is executed by the compartment that is allowing the enter to take place.

Once the dendritic cell has entered the lymph node, it can present the MHC class I molecules on its surface to nearby T cells. This activity

[3]http://www.ncbi.nlm.nih.gov/

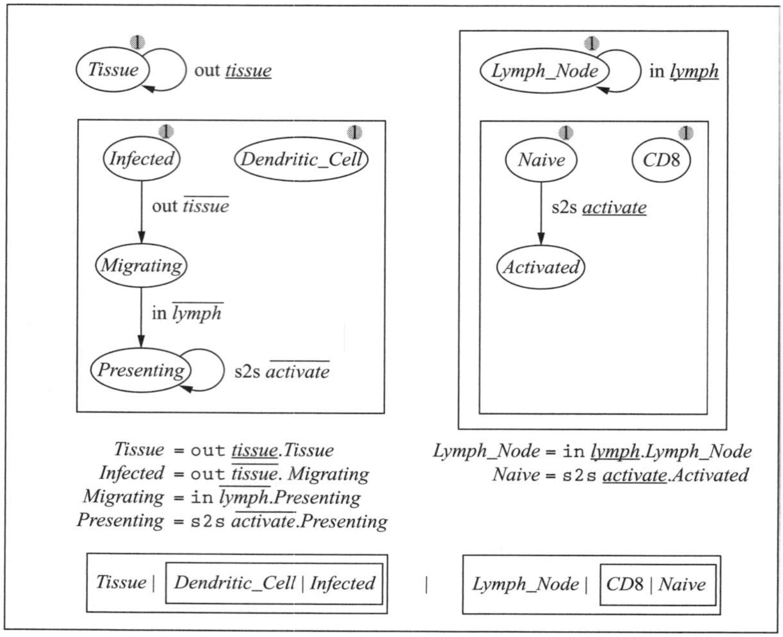

Figure 5.32 A SPiM model of some of the basic processes that can take place during an immune response to a viral infection

is represented by a persistent send on `s2s` $\overline{activate}$. The `s2s` keyword stands for *sibling to sibling*, and denotes a communication from one sibling compartment to another.

In general, the lymph nodes are meeting places where dendritic cells can present protein fragments of the viruses with which they are infected, and can recruit T cells to help fight the infection. Here we represent a CD8 T cell as a compartment, containing the $CD8$ process in parallel with the $Naive$ process. The $CD8$ process can represent a range of cellular processes that are necessary for the survival and normal functioning of the CD8 T cell. In our model, the $Naive$ process represents the current state of the CD8 T cell. The cell can transition from $Naive$ to *Activated* by receiving on `s2s` $\underline{activate}$ from a nearby cell. Here we represent the activation step as a single communication from the dendritic cell to the naive CD8 T cell on the *activate* channel. In reality, of course, the mechanism is much more complicated. In particular, the activation will be successful only if the MHC class I molecules and their contents on the surface of the dendritic cell match the receptors on the surface of the T cell. We can represent this match in an abstract way, by requiring the sender and receiver cells to communicate on the same *activate* channel. Once the T cell becomes activated, it can then leave the lymph node and circulate in the blood until it reaches the infected tissue; these later events are not depicted in Figure 5.32.

Figure 5.33 illustrates the sequence of events that take place when executing the model of Figure 5.32. For each frame in the figure, the graphical representation at the top is equivalent to the textual representation at the bottom. The number of copies of a given process is indicated next to the corresponding node in the figure. By default, we assume that a single copy

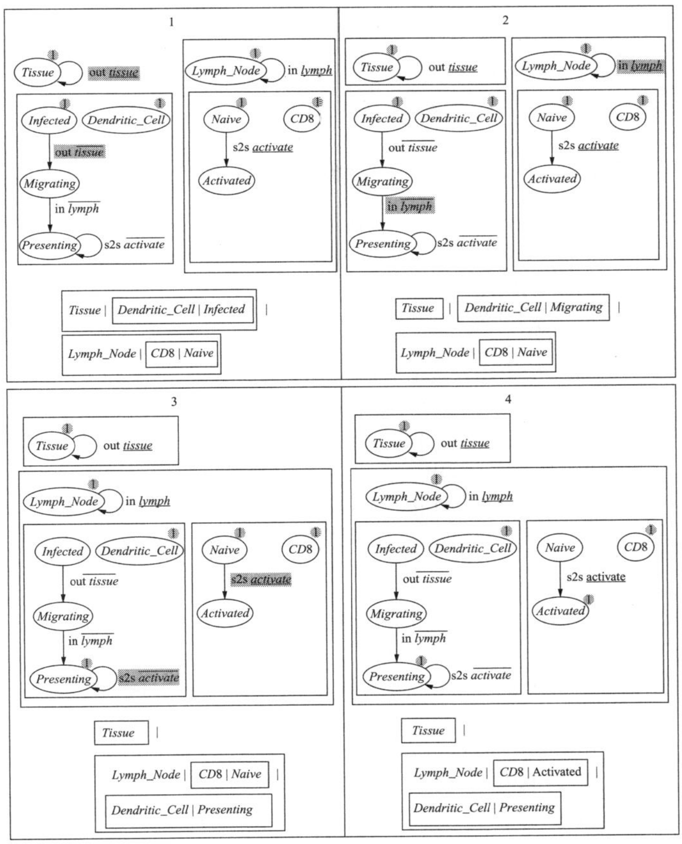

Figure 5.33 The sequence of events that take place when executing the model of Figure 5.32. (1) The infected dendritic cell can leave the tissue by sending on `out` $\overline{tissue}$. (2) The migrating dendritic cell has left the tissue, and can now enter the lymph node by sending on `in` $\overline{lymph}$. (3) The presenting dendritic cell can activate a nearby *Naive* CD8 T cell by sending on `s2s` $\overline{activate}$. (4) The CD8 T cell is now activated

of each compartment is present. The example illustrates the `in` and `out` movement primitives, which allow compartments to move in and out of one another, respectively. When a compartment moves it travels with its entire contents. When it reaches its new location, it can interact with additional compartments that may not have been present in its previous location. For example, in the case of the dendritic cell moving inside a lymph node, on arrival it can interact with *Naive* CD8 T cells that were not accessible in its previous location. The interaction between two compartments is achieved using the `s2s` communication primitive, which allows two sibling compartments to communicate with each other. Other primitives allow for communication from a parent to a child (`p2c`) and from a child to its parent (`c2p`); the `merge` primitive permits merging of two compartments. These primitives, together with the full definition of the SPiM calculus with compartments, are given in Section 5.7.5.

5.5.2 Simple Model of Viral Infection and T-Cell Response

Figure 5.34 extends the simple model of Figure 5.32 to include activation and proliferation of T cells. The model of a dendritic cell is extended to include an initial *Clean* process, which can receive on `s2s` *infect* from a nearby cell and become *Infected*. We also allow the *Presenting* process to *die*, which represents the fact that the dendritic cell will not continue presenting indefinitely. We start with a nominal population of 10 dendritic cells, where the initial population is indicated next to the compartment boundary. We provide a basic model for the other cells in the tissue, which start out in an uninfected state and can receive on `s2s` *infected*, becoming infected as a result. Once infected, the cell can then infect other neighboring cells by repeatedly sending on `s2s` *infect*. In this way, a virus inside a single cell can infect a potentially infinite number of neighboring cells. The infected cell can also be killed by receiving on `s2s` *kill*. We start with a nominal population of 1000 uninfected cells, together with a single infected cell.

We also extend the model of the tissue by allowing multiple cells to enter or leave, represented by persistent receive actions on `in` *tissue* and `out` *tissue*, respectively. Similarly, we allow multiple cells to enter or leave the lymph node, represented by persistent receive actions on `in` *lymph* and `out` *lymph*, respectively.

Finally, we extend the model of a CD8 T cell to represent the processes that take place after activation—namely, T-cell proliferation and migration to the site of infection, followed by the killing of the infected cells. Once activated, the CD8 T cell can either proliferate or die, which ensures that it does not continue to proliferate indefinitely. After the *proliferate* action is executed, a new cell is created in parallel with the *Parent* process. The creation of a new cell is represented by a triangle node, where the successor nodes to the triangle represent the contents of the newly created cell.

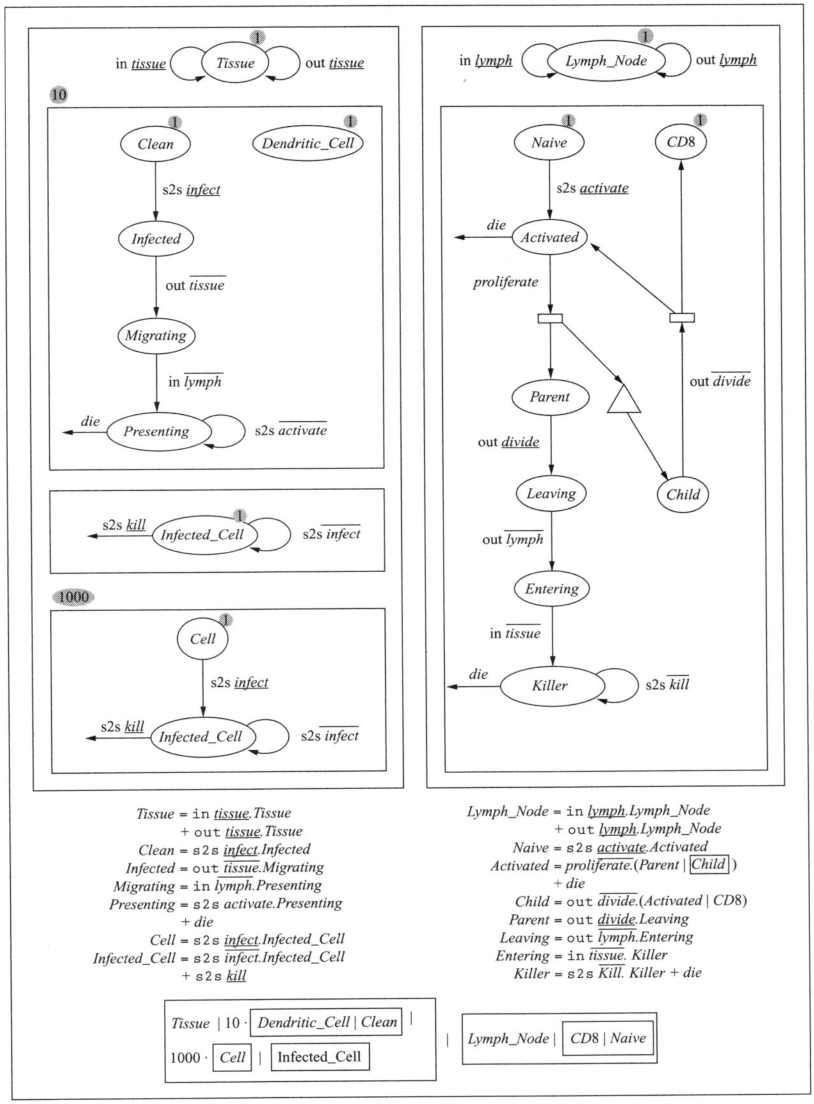

Figure 5.34 A simple model of viral infection and immune response

In this example, the new cell executes the *Child* process, which sends on out $\overline{divide}$. This is a simple way of representing cell division, in which the newly created cell leaves its parent and then executes alongside the parent. The *Parent* process allows the newly created cell to leave by receiving on `out` $\underline{divide}$. Once the child cell has left its parent, it executes the *Activated* and *CD8* processes, representing a newly created CD8 T cell in an activated state. Meanwhile, the parent cell can leave the lymph node by sending on `out` $\overline{lymph}$ and then enter the tissue by sending on `in` $\overline{tissue}$. On arrival, it can kill neighboring infected cells by repeatedly sending on `s2s` $\overline{kill}$, until eventually it dies. Note that the newly created T cell can perform the same actions as the activated parent. In this way, a single activated T cell can produce a large population of cells, thereby effectively fighting a viral infection.

Of course, this model is still an extreme simplification of the events that actually take place during T-cell activation. In particular, the activated CD8 T cell will be able to kill only those infected cells that it recognizes as being infected. As with T-cell activation, this determination will depend on whether the T-cell receptors recognize the viral protein fragments being presented on the surface of the infected cell by MHC class I molecules. We can model this complementarity between the presenter and the receiver by requiring the T cells and the infected cells to interact on the same *kill* channel. In this way, different signaling and receptor molecules can be represented by sending and receiving on different channels.

In addition, our simplified model represents proliferation by allowing a given cell to divide only once before leaving the lymph node; in reality, the same parent cell can divide multiple times. More generally, the lymph node has a much more complicated structure, and the behaviors of the individual cells will probably require millions of lines of code to be represented accurately. Nevertheless, we can continually refine our models to include additional detail as needed, and we can use our highly simplified models to study the dynamic interactions between cells at an abstract level.

Figure 5.35 shows a simulation of the model of Figure 5.34 (see color figures online). We start with 1000 tissue cells and a single infected tissue cell. In addition, 10 dendritic cells are assumed to be patrolling this particular area of tissue. When one of the dendritic cells becomes infected, it moves to a nearby lymph node, where it chances upon a *Naive* CD8 T cell that recognizes the viral protein fragments being presented on its surface and becomes activated as a result. The CD8 T cell proliferates and moves to the site of infection. Meanwhile, the infection is spreading between neighboring cells. We can see this trend by noting the decrease in the population of uninfected cells, shown in green, and the increase in the population of infected cells, shown in red. As the killer T cells arrive, they destroy the infected cells as a means of preventing the spread of the virus. We observe a sudden drop in the population of infected cells, which coincides with a rise in the population of killer T cells. Once the infection has been cleared, there are no more infected cells to stimulate an immune response, and the killer T cells die off.

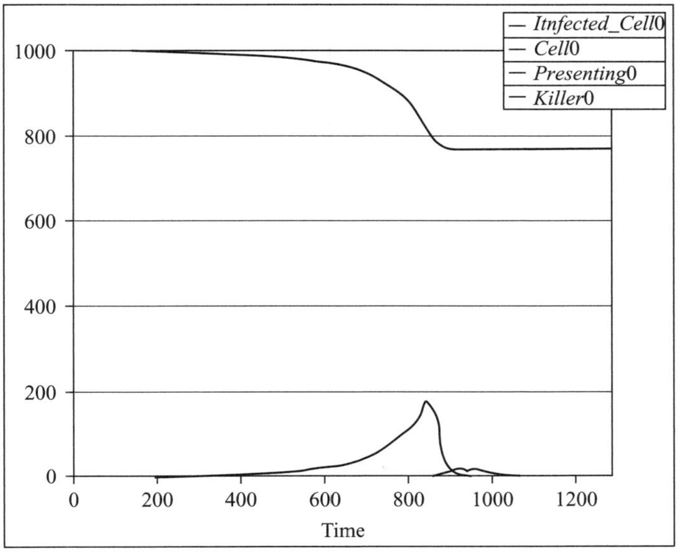

Figure 5.35 Simulation results for the viral infection model of Figure 5.34. The results show the total populations of uninfected and infected tissue cells, together with the populations of presenting dendritic cells and killer T cells over time. We assume that $proliferate = 0.5$, $divide = 0.01$, $activate = 0.2$, $killer = 0.005$, $infect = 0.00001$, and the remaining rates are equal to 1. We start with a single infected tissue cell among a population of 1000 uninfected cells and 10 dendritic cells. When a dendritic cell becomes infected, it moves to a lymph node, where it activates a corresponding CD8 T cell. The T cell then proliferates, with the newly produced T cells moving to the site of infection, where they kill the infected cells to contain the spread of the virus. After the infection is cleared, the T cells eventually die off

In working with this model, we can try different parameter values for the simulation to observe under which conditions the infection is contained and how many of the cells become infected. We can also observe the trade-offs between rapid proliferation with a large population of killer T cells and slower proliferation with fewer T cells. In reality, the immune system often tries to strike a balance between responding too strongly and using unnecessary resources versus not responding strongly enough and being unable to clear the infection. For the immune system to work both effectively and efficiently, the T cells need to proliferate at just the right rate depending on the level of presentation by the dendritic cells, and to quickly die off once the infection has been contained.

This model can be extended in a number of simple ways. For instance, rather than having a fixed population of tissue cells, we can include a steady-state population, in which cells are continually dying and proliferating. We can also include many other types of cells, such as helper T cells, which send activation signals to regulate the overall immune response. Furthermore, when a dendritic cell becomes infected by a virus such as measles, the dendritic cells can also cause the virus to spread to other cells in the lymph node, leading to a sophisticated battle between immune system and virus.

5.6 Conclusion

This chapter presented a visual process calculus for designing and simulating computer models of biological systems. We started with simple models of interacting genes and proteins, then introduced more complex models of gene networks. We also presented simple models of developmental processes by constructing a generic model of a particular cell type, then instantiating this model to represent a number of identical cells in different locations with different environmental signals. Finally, we used the calculus to model cells that can move, proliferate, and interact with one another in different physical and logical locations.

The calculus presented here builds on previous work [28, 29, 31, 32], but there is still scope for many extensions—for example, to handle spatiotemporal dynamics or provide for more realistic biological interactions. One example is the use of primitives for biological membranes [4], which more accurately represent membrane formation and interaction. More generally, a wide variety of process calculi are being developed for biology [7, 8, 10, 23, 27]. In the future, these and other process calculi could help to form the theoretical foundations for novel programming languages for biology. Such languages will enable us to take modeling, simulation, and analysis of biological systems to a stage where we can handle models consisting of millions of lines of code in a scalable, modular fashion, just as we have achieved for more traditional software systems.

5.7 The SPiM Calculus

This section presents the technical definition of the SPiM calculus, a variant of stochastic pi-calculus for biological modeling.

5.7.1 Syntax

The syntax of the calculus is presented in Definition 5.1, where r denotes a *rate* belonging to the set of real numbers, x denotes a *channel,* X denotes a *species,* $\tilde{n}$ denotes a *list* of names $n_1, \ldots, n_N$, and $\tilde{m}$ denotes a list of names that are pairwise distinct. Each channel x is also associated with

Definition 5.1 Syntax of SPiM.

$\pi ::=$	r	Delay at rate r
$\mid$	$\underline{x}(\tilde{m})^r$	Receive names $\tilde{m}$ on channel x at rate r
$\mid$	$\overline{x}\langle\tilde{n}\rangle^r$	Send names $\tilde{n}$ on channel x at rate r
$\mid$	$\overline{x}(\tilde{m})^r$	Send restricted names $\tilde{m}$ on channel x at rate r
$M ::=$	$\pi_1.P_1 + \ldots + \pi_N.P_N$	Choice between actions, $N \geq 0$
$P ::=$	$\mathbf{0}$	Null
$\mid$	$X(\tilde{n})$	Species X with parameters $\tilde{n}$
$\mid$	$P_1 \mid \ldots \mid P_N$	Parallel composition of processes, $N \geq 2$
$\mid$	$(x_1) \ldots (x_N)\, P$	Restricted channels $x_1, \ldots,$ x_N in P
$D ::=$	P	Definition of a process
$\mid$	M	Definition of a choice
$E ::=$	$X_1(\tilde{m}_1){=}D_1, \ldots, X_N(\tilde{m}_N){=}D_N$	Definition of species X_i with parameters $\tilde{m}_i$
$S ::=$	$E \vdash P$	System with environment E and process P

a corresponding rate given by $\rho(x)$. Rates are used to characterize the speed of a reaction according to an exponential distribution, such that the probability of a reaction with rate r occurring within time t is given by $F(t) = 1 - e^{-r \cdot t}$. The average duration of the reaction is given by the mean $1/r$ of this distribution.

An *action* π can be a delay at rate r, a receive $\underline{x}(\tilde{m})^r$ of names $\tilde{m}$ on channel x at rate r, a send $\overline{x}\langle\tilde{n}\rangle^r$ of names $\tilde{n}$ on channel x at rate r, or a send $\overline{x}(\tilde{m})^r$ of restricted names $\tilde{m}$ on channel x at rate r. A choice M denotes a competition between zero or more actions of the form $\pi.P$, where process P is executed after performing action π. A process P can be an empty process $\mathbf{0}$, a species $X(\tilde{n})$ with name X and parameters $\tilde{n}$, a parallel composition of two or more processes $P_1 \mid \ldots \mid P_N$, or a process $(x_1) \ldots (x_N)\, P$ with restricted channels $x_1, \ldots, x_N$. The restricted channels are used to represent the formation of a complex. For example, the process $(x)\,(X_1(x) \mid X_2(x))$ represents a complex of two species, X_1 and X_2, bound to each other on channel x. The restriction denotes a local channel x on which the two species can interact to split the complex.

An environment E consists of a set of species definitions of the form $X(\tilde{m}) = D$, where X is the name of the species, $\tilde{m}$ are its parameters, and D is its corresponding definition, which can be a process P or a choice M. It is assumed that $\mathrm{fn}(D) \subseteq \tilde{m}$, where $\mathrm{fn}(D)$ denotes the set of free names of D, given that $(x)\,P$ binds the name x in P, and $\underline{x}(\tilde{m})^r.P$ and $\overline{x}(\tilde{m})^r.P$ bind the set of names $\tilde{m}$ in P. Definitions D are also assumed to be equal up to renaming of bound names. In addition, for each definition $X(\tilde{m}) = D$, any calls to X inside D can occur only after an action π. This constraint prevents the existence of potentially infinite processes such as $X() = (X() \mid X())$. Finally, a system S consists of a global constant environment E, together with a process P.

We define a number of syntactic conventions: we assume that the parallel composition operator ($\mid$) has the lowest precedence among the operators of the calculus; we abbreviate an action $\pi.\mathbf{0}$ followed by the empty process to π, and a sequence of restricted channels $(x_1) \ldots (x_N)\, P$ to $(x_1, \ldots, x_N)\, P$; we abbreviate a parallel composition of N identical processes $P \mid \ldots \mid P$ to $N \cdot P$. We also allow some or all of the parameters of a species to be hidden in cases where these parameters are unchanged throughout the system. We can recover the parameters by observing the free names of the species definition. To model the binding and unbinding of species, we assume that each channel x is associated with a unique *unbinding channel* ^-x, and vice versa, where x and ^-x represent two distinct names. We write $^+\overline{x}$ as an abbreviation for $\overline{x}(^-x)$ and $^+\underline{x}$ as an abbreviation for $\underline{x}(^-x)$. We illustrate these abbreviations with the following three identical systems, which model the binding and unbinding of proteins X and Y, respectively, as described in Figure 5.9:

$$\begin{array}{ll} X = {}^+\overline{b}.X', & Y = {}^+\underline{b}.Y', \\ X' = {}^-\overline{b}.X, & Y' = {}^-\underline{b}.Y \end{array} \quad \vdash X \mid Y \tag{5.1}$$

$$\begin{array}{ll} X = \overline{b}(^-b).X', & Y = \underline{b}(^-b).Y', \\ X' = {}^-\overline{b}.X, & Y' = {}^-\underline{b}.Y \end{array} \quad \vdash X \mid Y \tag{5.2}$$

$$\begin{array}{ll} X(b) = \overline{b}(^-b).X'(b, {}^-b), & Y(b) = \underline{b}(^-b).Y'(b, {}^-b), \\ X'(b, {}^-b) = {}^-\overline{b}.X(b), & Y'(b, {}^-b) = {}^-\underline{b}.Y(b) \end{array} \quad \vdash X(b) \mid Y(b) \tag{5.3}$$

The first system models the binding and unbinding of proteins X and Y, respectively, on channel b, using the syntactic abbreviations for binding. The second system expands the binding abbreviations. The third system explicitly shows all of the parameters of the different proteins. The abbreviations allow a more compact representation of systems, without any loss of information. Note that here we have defined binding and unbinding primitives purely by means of syntactic abbreviations. A more detailed treatment of such primitives is given in [6], which defines a novel calculus for binding and unbinding.

5.7.2 Reaction

The reaction rules of the calculus are presented in Definition 5.2. The rules are of the form $\frac{condition}{reaction}$, where the given *condition* must be satisfied for the given *reaction* to be possible. The notation $P \xrightarrow{r} P'$ states that the process P can evolve to P' by performing a reaction at rate r. If a species $X(\tilde{n})$ can do a delay $r.P$ in competition with actions M, then the species can evolve to a process P at rate r (5.6). If a species $X_1(\tilde{n}_1)$ can do a send $\overline{x}\langle\tilde{n}\rangle^{r_1}.P_1$ in competition with actions M_1, and in parallel a species $X_2(\tilde{n}_2)$ can do a receive $\underline{x}(\tilde{m})^{r_2}.P_2$ in competition with actions M_2, then the two species can interact at the rate of the channel x times the rates r_1 and r_2 of the actions. After the interaction takes place, the processes P_1 and P_2 are

Definition 5.2 Reaction rules in SPiM. The condition $X(\tilde{n}) = D'$ is true if there is a definition in the system environment such that $X(\tilde{m}) = D$ and $D' = D_{\{\tilde{n}/\tilde{m}\}}$.

$$\frac{X_1(\tilde{n}_1) = \overline{x}\langle\tilde{n}\rangle^{r_1}.P_1 + M_1 \quad X_2(\tilde{n}_2) = \underline{x}(\tilde{m})^{r_2}.P_2 + M_2}{X_1(\tilde{n}_1) \,|\, X_2(\tilde{n}_2) \xrightarrow{\rho(x)\cdot r_1 \cdot r_2} P_1 \,|\, P_{2\{\tilde{n}/\tilde{m}\}}} \tag{5.4}$$

$$\frac{X_1(\tilde{n}_1) = \overline{x}(\tilde{m})^{r_1}.P_1 + M_1 \quad X_2(\tilde{n}_2) = \underline{x}(\tilde{m})^{r_2}.P_2 + M_2}{X_1(\tilde{n}_1) \,|\, X_2(\tilde{n}_2) \xrightarrow{\rho(x)\cdot r_1 \cdot r_2} (\tilde{m})(P_1 \,|\, P_2)} \tag{5.5}$$

$$\frac{X(\tilde{n}) = r.P + M}{X(\tilde{n}) \xrightarrow{r} P} \tag{5.6}$$

$$\frac{P \xrightarrow{r} P'}{(x)\,P \xrightarrow{r} (x)\,P'} \tag{5.7}$$

$$\frac{P \xrightarrow{r} P'}{P \,|\, Q \xrightarrow{r} P' \,|\, Q} \tag{5.8}$$

$$\frac{Q \equiv P \xrightarrow{r} P' \equiv Q'}{Q \xrightarrow{r} Q'} \tag{5.9}$$

Definition 5.3 Structural congruence axioms in SPiM. *Structural congruence is reflexive, symmetric, and transitive, and holds in any context inside a process or a choice.*

$$P \mid \mathbf{0} \equiv P \quad (5.10)$$
$$P_1 \mid P_2 \equiv P_2 \mid P_1 \quad (5.11)$$
$$P_1 \mid (P_2 \mid P_3) \equiv (P_1 \mid P_2) \mid P_3 \quad (5.12)$$
$$\pi_1.P_1 + \pi_2.P_2 \equiv \pi_2.P_2 + \pi_1.P_1 \quad (5.13)$$
$$(x)\,\mathbf{0} \equiv \mathbf{0} \quad (5.14)$$
$$(x)\,(y)\,P \equiv (y)\,(x)\,P \quad (5.15)$$
$$(x)\,(P_1|P_2) \equiv P_1|\,(x)P_2 \,\textbf{if}\; x \notin \text{fn}(P_1) \quad (5.16)$$
$$X(\tilde{n}) \equiv P_{\{\tilde{n}/\tilde{m}\}} \,\textbf{if}\, X(\tilde{m}) = P \quad (5.17)$$

executed in parallel, where the parameters $\tilde{m}$ are replaced with $\tilde{n}$ in process P_2, written $P_{2\,\{\tilde{n}/\tilde{m}\}}$ (5.4). If the species $X_1(\tilde{n}_1)$ can do a restricted send $\overline{x}(\tilde{m})^{r_1}.P_1$ in competition with actions M_1, then the two species become bound on channels $\tilde{m}$ after the interaction takes place, and the result of the interaction is the complex $(\tilde{m})\,(P_1 \mid P_2)$ (5.5). All of these reactions can also take place inside a restriction (5.7), inside a parallel composition (5.8), and up to reordering of processes (5.9). The reordering is defined by structural rules, according to Definition 5.3. Essentially, the rules allow the reordering of parallel processes (5.11), actions (5.13), and restricted channels (5.15), and the removal of empty parallel processes (5.10) and unused restricted channels (5.14). Parallel composition is associative (5.13), and the scope of a restricted channel can be extended over parallel processes that do not use the channel (5.16). Finally, a species $X(\tilde{n})$ defined in the environment as $X(\tilde{m}) = P$ can be replaced with its process definition, where the parameters $\tilde{m}$ are replaced with $\tilde{n}$ in process P (5.17).

5.7.3 Reaction Probability

According to standard principles of chemical kinetics [13], the probability of a reaction should be proportional to its rate. To compute this probability, we need a way of computing the rates of each of the individual reactions in the system. We do so by using a notion of standard form, presented in Definition 5.4, together with a set of indexed reaction rules, presented in Definition 5.5. This approach is inspired by previous work [16]. A given process can be converted to standard form by application of the structural

Definition 5.4 A process P is in *standard form* if it is in the form given by (5.18), such that all restricted channels are at the top level, and each species $X_i(n_i)$ is defined as a choice of actions M_i in the environment.

$$(x_1) \ldots (x_M)\,(X_1(\tilde{n}_1) \mid \ldots \mid X_N(\tilde{n}_N)) \tag{5.18}$$

Definition 5.5 Indexed reaction rules in SPiM. *Processes are assumed to be in standard form, in accordance with Definition 5.4. A choice between zero or more actions* $\pi_1.P_1 + \ldots + \pi_N.P_N$ *is abbreviated to a sum* $\sum_{i\in\{1\ldots N\}} \pi_i.P_i$ *and a parallel composition of zero or more processes* $P_1 \mid \ldots \mid P_N$ *is abbreviated to a product* $\prod_{i\in\{1\ldots N\}} P_i$.

$$\frac{\begin{array}{llll} X_{i_1}(\tilde{n}_{i_1}) = \sum_{j\in J_1} \pi_j.P_j & \pi_{j_1} = \overline{x}\langle\tilde{n}\rangle^{r_1} & j_1 \in J_1 & \\ X_{i_2}(\tilde{n}_{i_2}) = \sum_{j\in J_2} \pi_j.P_j & \pi_{j_2} = \underline{x}(\tilde{m})^{r_2} & j_2 \in J_2 & i_1 \neq i_2 \quad i_1, i_2 \in I \end{array}}{\prod_{i\in I} X_i(\tilde{n}_i) \xrightarrow{(\rho(x)\cdot r_1\cdot r_2),(i_1,j_1,i_2,j_2)} P_{j_1} \mid P_{j_2\,\{\tilde{n}/\tilde{m}\}} \mid \prod_{i\in I\setminus i_1,i_2} X_i(\tilde{n}_i)} \tag{5.19}$$

$$\frac{\begin{array}{llll} X_{i_1}(\tilde{n}_{i_1}) = \sum_{j\in J_1} \pi_j.P_j & \pi_{j_1} = \overline{x}(\tilde{m})^{r_1} & j_1 \in J_1 & \\ X_{i_2}(\tilde{n}_{i_2}) = \sum_{j\in J_2} \pi_j.P_j & \pi_{j_2} = \underline{x}(\tilde{m})^{r_2} & j_2 \in J_2 & i_1 \neq i_2 \quad i_1, i_2 \in I \end{array}}{\prod_{i\in I} X_i(\tilde{n}_i) \xrightarrow{(\rho(x)\cdot r_1\cdot r_2),(i_1,j_1,i_2,j_2)} (\tilde{m})\,(P_{j_1} \mid P_{j_2}) \mid \prod_{i\in I\setminus i_1,i_2} X_i(\tilde{n}_i)} \tag{5.20}$$

$$\frac{X_{i_1}(\tilde{n}_{i_1}) = \sum_{j\in J} \pi_j.P_j \qquad \pi_{j_1} = r \qquad i_1 \in I, j_1 \in J}{\prod_{i\in I} X_i(\tilde{n}_i) \xrightarrow{r,(i_1,j_1)} P_{j_1} \mid \prod_{i\in I\setminus i_1} X_i(\tilde{n}_i)} \tag{5.21}$$

$$\frac{P \xrightarrow{r,w} P'}{(x)\,P \xrightarrow{r,w} (x)\,P'} \tag{5.22}$$

$$\frac{P \xrightarrow{r,w} P' \equiv Q'}{P \xrightarrow{r,w} Q'} \tag{5.23}$$

rules in Definition 5.3. In particular, all of the restricted channels can be moved to the top level by application of rule (5.16), and each of the species defined as a process can be expanded by application of rule (5.17), until only the species defined as a choice of actions remain.

Once a process is in standard form, we can identify each species $X_i(n_i)$ by its position i in the parallel composition $X_1(n_1) \mid \ldots \mid X_N(n_N)$, and we can identify each action $\pi_j.P_j$ of a given species by its position j inside the choice $\pi_1.P_1 + \ldots + \pi_M.P_M$. Thus each action can be identified by a pair (i, j). These action identifiers can then be used to identify all of the reactions in a system. A reaction identifier w can be a pair of indices (i, j) denoting a delay action j inside choice i, or a tuple of four indices (i_1, j_1, i_2, j_2) denoting a send action j_1 inside choice i_1 interacting with a receive action j_2 inside choice i_2. The indexed reaction rules of Definition 5.5 are analogous to those of Definition 5.2, except that they assign an identifier w to each reaction, based on the position of each action in the system. The notation $P \xrightarrow{r,w} P'$ states that the process P can evolve to P' by performing a reaction with identifier w at rate r.

We can use the indexed reaction rules to compute the probability of individual reactions, as shown in Definition 5.6. The propensity $\rho(P)$ of a process P is defined as the sum of the rates of all reactions of the process (5.24). The probability $\Pr(P \xrightarrow{r,w} P')$ that the process P can perform a particular reaction w with rate r and evolve to P' is given by the rate of the reaction divided by the propensity of the process (5.25). Similarly, the probability $\Pr(P \longrightarrow P')$ that the process P can reduce to P' is given by the sum of the probabilities of all reactions for which P can reduce to P' (5.26). These reaction probabilities form the basis of the SPiM stochastic simulation algorithm. More details about the algorithm and its corresponding implementation are presented in [25].

Definition 5.6 Reaction Probability in SPiM. Equation (5.24) defines the propensity of process P, where $P \xrightarrow{r,w}$ means that P can perform a reaction w at rate r. Equation (5.25) defines the probability that process P can perform a particular reaction w at rate r and evolve to P'. Equation (5.26) defines the probability that process P can evolve to P'.

$$\rho(P) \triangleq \sum_{P \xrightarrow{r,w}} r \tag{5.24}$$

$$\Pr(P \xrightarrow{r,w} P') \triangleq \text{if } P \xrightarrow{r,w} P' \text{ then } \frac{r}{\rho(P)} \text{ else } 0 \tag{5.25}$$

$$\Pr(P \longrightarrow P') \triangleq \sum_{P \xrightarrow{r,w} P'} \frac{r}{\rho(P)} \tag{5.26}$$

5.7.4 Graphical Representation

The syntax of the SPiM calculus supports a concise graphical representation, originally presented in [26]. A system $E \vdash P$ is displayed in two parts: a static environment E that remains constant over time, and a dynamic process P that is updated after each reaction.

The static environment E is displayed as a *graph* of *nodes* and *edges*, as shown in Figure 5.36. Each choice M or process P in the environment is displayed as a node in the graph, and each definition $X(\tilde{m}) = D$ assigns an *identifier* X and a *label* $X(\tilde{m})$ to a given node D. A choice $\pi_1.P_1 + \ldots + \pi_N.P_N$ is displayed as an elliptical node with edges to nodes $P_1, \ldots, P_N$. Each edge to a node P_i is labeled with an action π_i and denotes an alternative execution path in the system. A parallel composition $P_1 \mid \ldots \mid P_N$ is displayed as a rectangular node with edges to nodes $P_1, \ldots, P_N$. Each edge to a node P_i denotes a concurrent execution path in the system. A species $X(\tilde{n})$ with definition $X(\tilde{m}) = D$ is displayed as a rectangular node with an edge to the node identified by X. If $\tilde{m} \neq \tilde{n}$, then the tip of the edge is labeled with the *substitution* $\tilde{m} := \tilde{n}$. This expression represents the passing of parameters from one node to another. Note that we need to display only those mappings in the substitution that are different from the identity, as we need to indicate only those parameters that are modified by the substitution. We also abbreviate the notation so that edges from a choice or parallel composition node to a species $X(\tilde{n})$ are connected directly to node X, without displaying the additional rectangle. For example, an action $\pi.X(\tilde{n})$ with $X(\tilde{n}) = D$ is displayed as $\bigcirc\!\xrightarrow{\pi}\!X$ instead of $\bigcirc\!\xrightarrow{\pi}\!\square\!\longrightarrow\!X$. A restriction $(\tilde{m})\,P$ is displayed by placing the restricted channels $(\tilde{m})$ next to node P, and the null process $\mathbf{0}$ is not displayed.

To identify the dynamic process P, we attach *tokens* to nodes in the graph to represent currently executing species, and then connect restricted

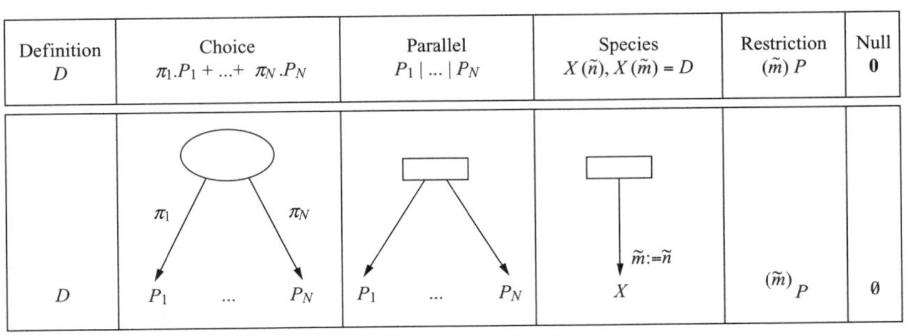

Figure 5.36 Graphical representation of a static environment E in SPiM, where the graphical representation at the bottom is equivalent to the textual representation at the top

names to the tokens to represent the formation of *complexes*, as shown in Figure 5.37. A species $X(\tilde{n})$ with definition $X(\tilde{m}) = D$ is displayed by attaching a substitution token $\tilde{m} := \tilde{n}$ to the node identified by X. A parallel composition of species $X_1(\tilde{n}_1) \mid \ldots \mid X_N(\tilde{n}_N)$ that share restricted names $(\tilde{m})$ is displayed by drawing a dotted edge from names $\tilde{m}$ to the tokens of each of the species, thereby indicating the formation of a complex of species. The null process $\mathbf{0}$ is not displayed.

We introduce a number of convenient abbreviations for the graphical representation of dynamic processes. If N identical tokens are attached to the same node, they can be replaced with a single token preceded by the number N. Furthermore, if $\tilde{m} = \tilde{n}$, the token can be displayed as just the number N. If there are N copies of a restriction $(\tilde{m})\,P$, they can be displayed by placing the number N next to the restricted names $\tilde{m}$. The scope of a restricted name can be further clarified by drawing a dotted edge to species that use this name. For example, a restriction $(x_1, x_2)\,(X_1(x_1) \mid X_2(x_1, x_2) \mid X_3(x_2))$ can be displayed as $X_1{(1)} \cdots x_1 \cdots {(1)}X_2{(1)} \cdots x_2 \cdots {(1)}X_3$, assuming that the substitution tokens are all empty and displayed as (1). The graphical representation clearly shows that X_1 and X_3 do not share any restricted names and, therefore, are not connected to each other directly. Finally, in some cases a graph can appear crowded when multiple tokens with different substitutions are attached to the same node. To improve readability in these cases, a separate copy of the entire graph may be displayed for each of the different tokens.

Once we have formally defined a graphical syntax in this way, we can use it to display the execution of a model. Essentially, the execution of a system can be visualized by applying the reaction rules to a given process in SPiM, and then displaying the resulting process after each reaction. Thus we do not need to formally define the graphical reaction rules. We illustrate

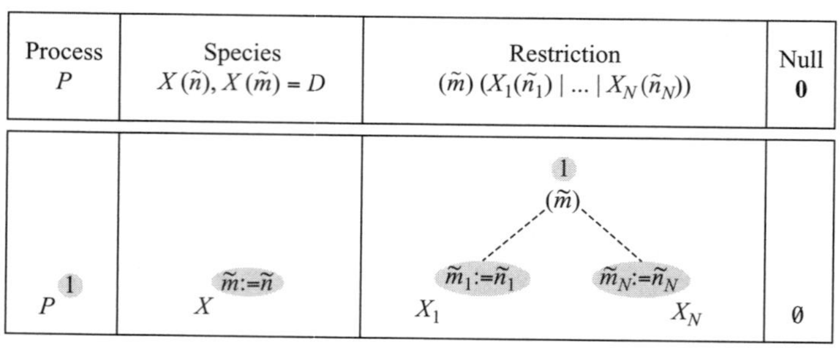

Figure 5.37 Graphical representation of a dynamic process P in SPiM, where the graphical representation at the bottom is equivalent to the textual representation at the top

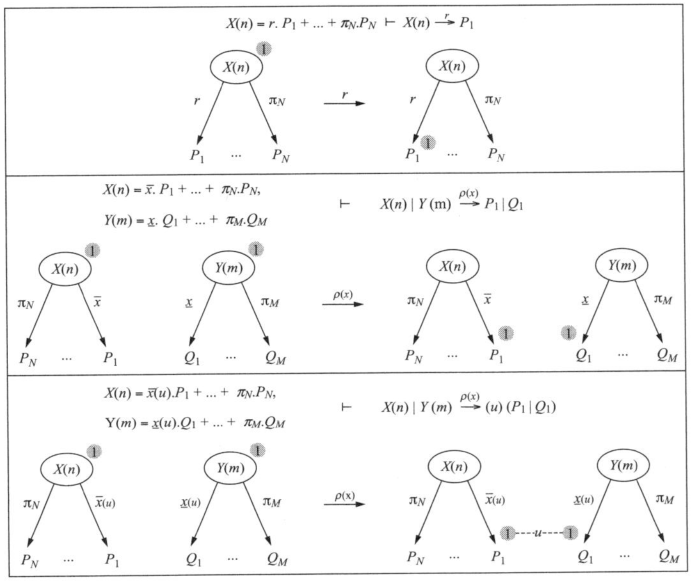

Figure 5.38 Example graphical reactions in SPiM, where the graphical representation at the bottom is equivalent to the textual representation at the top. The examples illustrate the three main reaction rules (5.6), (5.4), and (5.5) of Definition 5.2, respectively. A given process is executed graphically by first applying one of the calculus reaction rules and then displaying the resulting process

a number of example reactions by displaying the process before and after the reaction takes place, as shown in Figure 5.38.

5.7.5 SPiM with Compartments

The SPiM calculus can be extended with mobile compartments, resulting in a calculus similar to the bioambient calculus originally presented in [31]. The syntax of SPiM with compartments is a direct extension of Definition 5.7. The syntax of processes is extended with the notion of an *ambient* $\boxed{P}$ consisting of a process P inside a compartment. The syntax of actions is extended so that an action π can also be a receive $\gamma\underline{x}(\tilde{m})$ of names $\tilde{m}$ on channel x from an ambient, a send $\gamma\overline{x}\langle\tilde{n}\rangle$ of names $\tilde{n}$ on channel x to an ambient, or a send $\gamma\overline{x}(\tilde{m})$ of restricted names $\tilde{m}$ on channel x to an ambient, where γ denotes the direction of communication. This communication can occur from a child ambient to its parent (`c2p`), from a parent ambient to a child (`p2c`), or from one sibling ambient to another (`s2s`). In addition, an action π can be a move $\mu\overline{x}$ on channel x or an accept $\mu\underline{x}$ on channel x, where μ denotes the direction of movement. This movement

Definition 5.7 Syntax of SPiM with compartments, which extends the syntax of Definition 5.1, where γ represents a direction of communication and μ represents a direction of movement.

$P, Q ::=$	$\ldots$		$\gamma ::=$	`s2s`	Sibling
$\mid$	$\boxed{P}$	Ambient containing process P	$\mid$	`c2p`	Parent
			$\mid$	`p2c`	Child
$\pi ::=$	$\ldots$				
$\mid$	$\gamma\underline{x}(\tilde{m})$	Ambient receive	$\mu ::=$	`in`	Enter
$\mid$	$\gamma\overline{x}\langle\tilde{n}\rangle$	Ambient send	$\mid$	`out`	Leave
$\mid$	$\gamma\overline{x}(\tilde{n})$	Ambient restricted send	$\mid$	`merge`	Merge
$\mid$	$\mu\overline{x}$	Ambient move			
$\mid$	$\mu\underline{x}$	Ambient accept			

may consist of an ambient entering one of its siblings (`in`), a child ambient leaving its parent (`out`), or the merging of two sibling ambients (`merge`).

The reaction rules for SPiM with compartments are presented in Definition 5.8, and extend the reaction rules of Definition 5.2. The rules (5.27), (5.28), and (5.29) allow two ambients to communicate with each other. If a species $X_1(\tilde{n}_1)$ can do a send $\gamma\overline{x}\langle\tilde{n}\rangle^{r_1}.P_1$ in competition with actions M_1, and in parallel a species $X_2(\tilde{n}_2)$ can do a receive $\gamma\underline{x}(\tilde{m})^{r_2}.P_2$ in competition with actions M_2, then the two species can interact at the rate of the channel x times the rates r_1 and r_2 of the actions. After the interaction takes place, process P_1 continues executing in the sending ambient, and process P_2 continues executing in the receiving ambient, where the parameters $\tilde{m}$ are replaced with the names $\tilde{n}$ in process P_2, written $P_{2\,\{\tilde{n}/\tilde{m}\}}$. The communication can be from a child to a parent (5.27), from a parent to a child (5.28) or from one sibling to another (5.29).

The rules (5.30), (5.31), and (5.32) allow two ambients to move in or out of each other, or to merge with each other. If a species $X_1(\tilde{n}_1)$ can do a move $\mu\overline{x}^{r_1}.P_1$ in competition with actions M_1, and in parallel a species $X_2(\tilde{n}_2)$ can do an accept $\mu\underline{x}^{r_2}.P_2$ in competition with actions M_2, then the two species can interact at the rate of the channel x times the rates r_1 and r_2 of the actions. After the interaction takes place, process P_1 continues executing in the moving ambient, and process P_2 continues executing in the accepting ambient. The move can be one sibling entering another (5.30),

Definition 5.8 Reaction rules for SPiM with compartments, which extend the reaction rules of Definition 5.2. The condition $X(\tilde{n}) = D'$ is true if there is a definition in the system environment such that $X(\tilde{m}) = D$ and $D' = D_{\{\tilde{n}/\tilde{m}\}}$. For each of the communication rules (5.27), (5.28), and (5.29) there is also a corresponding rule for restricted communication, along the lines of (5.4) (not shown).

$$\frac{X_1(\tilde{n}_1) = \mathtt{c2p}\,\overline{x}\langle\tilde{n}\rangle^{r_1}.P_1 + M_1 \qquad X_2(\tilde{n}_2) = \mathtt{c2p}\,\underline{x}(\tilde{m})^{r_2}.P_2 + M_2}{\boxed{Q_1 \mid X_1(\tilde{n}_1)} \mid Q_2 \mid X_2(\tilde{n}_2) \xrightarrow{\rho(x)\cdot r_1\cdot r_2} \boxed{Q_1 \mid P_1} \mid Q_2 \mid P_{2\{\tilde{n}/\tilde{m}\}}} \tag{5.27}$$

$$\frac{X_1(\tilde{n}_1) = \mathtt{p2c}\,\overline{x}\langle\tilde{n}\rangle^{r_1}.P_1 + M_1 \qquad X_2(\tilde{n}_2) = \mathtt{p2c}\,\underline{x}(\tilde{m})^{r_2}.P_2 + M_2}{Q_1 \mid X_1(\tilde{n}_1) \mid \boxed{Q_2 \mid X_2(\tilde{n}_2)} \xrightarrow{\rho(x)\cdot r_1\cdot r_2} Q_1 \mid P_1 \mid \boxed{Q_2 \mid P_{2\{\tilde{n}/\tilde{m}\}}}} \tag{5.28}$$

$$\frac{X_1(\tilde{n}_1) = \mathtt{s2s}\,\overline{x}\langle\tilde{n}\rangle^{r_1}.P_1 + M_1 \qquad X_2(\tilde{n}_2) = \mathtt{s2s}\,\underline{x}(\tilde{m})^{r_2}.P_2 + M_2}{\boxed{Q_1 \mid X_1(\tilde{n}_1)} \mid \boxed{Q_2 \mid X_2(\tilde{n}_2)} \xrightarrow{\rho(x)\cdot r_1\cdot r_2} \boxed{Q_1 \mid P_1} \mid \boxed{Q_2 \mid P_{2\{\tilde{n}/\tilde{m}\}}}} \tag{5.29}$$

$$\frac{X_1(\tilde{n}_1) = \mathtt{in}\,\overline{x}^{r_1}.P_1 + M_1 \qquad X_2(\tilde{n}_2) = \mathtt{in}\,\underline{x}^{r_2}.P_2 + M_2}{\boxed{Q_1 \mid X_1(\tilde{n}_1)} \mid \boxed{Q_2 \mid X_2(\tilde{n}_2)} \xrightarrow{\rho(x)\cdot r_1\cdot r_2} \boxed{\boxed{Q_1 \mid P_1} \mid Q_2 \mid P_2}} \tag{5.30}$$

$$\frac{X_1(\tilde{n}_1) = \mathtt{out}\,\overline{x}^{r_1}.P_1 + M_1 \qquad X_2(\tilde{n}_2) = \mathtt{out}\,\underline{x}^{r_2}.P_2 + M_2}{\boxed{\boxed{Q_1 \mid X_1(\tilde{n}_1)} \mid Q_2 \mid X_2(\tilde{n}_2)} \xrightarrow{\rho(x)\cdot r_1\cdot r_2} \boxed{Q_1 \mid P_1} \mid \boxed{Q_2 \mid P_2}} \tag{5.31}$$

$$\frac{X_1(\tilde{n}_1) = \mathtt{merge}\,\overline{x}^{r_1}.P_1 + M_1 \qquad X_2(\tilde{n}_2) = \mathtt{merge}\,\underline{x}^{r_2}.P_2 + M_2}{\boxed{Q_1 \mid X_1(\tilde{n}_1)} \mid \boxed{Q_2 \mid X_2(\tilde{n}_2)} \xrightarrow{\rho(x)\cdot r_1\cdot r_2} \boxed{Q_1 \mid P_1 \mid Q_2 \mid P_2}} \tag{5.32}$$

$$\frac{P \xrightarrow{r} P'}{\boxed{P} \xrightarrow{r} \boxed{P'}} \tag{5.33}$$

---

a child leaving its parent (5.31), or two siblings merging (5.32). In the case of the merge, the contents of both ambients are merged in a single ambient. In addition, all of these reactions can take place inside an ambient (5.33).

The structural rules for SPiM with compartments are extended with the rules in Definition 5.9, which allow empty ambients to be deleted (5.34) and

Definition 5.9 Structural congruence axioms for SPiM with compartments, which extend the axioms of Definition 5.2.

$$\boxed{\mathbf{0}} \equiv \mathbf{0} \tag{5.34}$$

$$(x)\,\boxed{P} \equiv \boxed{(x)\,P} \tag{5.35}$$

---

the scope of restricted names to be extended outside an ambient (5.35). The reaction probabilities for SPiM with compartments can be derived in a similar way to Definitions 5.5 and 5.6, although we omit the details here. More information about a stochastic simulation algorithm for bioambients and its corresponding implementation can be found in [24].

The SPiM calculus with compartments also supports a concise graphical representation, as shown in Figure 5.39. In a static environment, an ambient $\boxed{P}$ is displayed as a triangular node with an edge to process P. In a dynamic process, the ambient is displayed as a box around the process P, where each box contains its own copy of the graph representing P. If N identical ambients are executing in parallel, only one copy of the ambient is represented, with the number N next to the surrounding box. Once again, the graphical reaction rules do not need to be defined explicitly. Instead, the execution of a system can be visualized by displaying the system before and after each reaction. Figure 5.40 illustrates a number of example reactions.

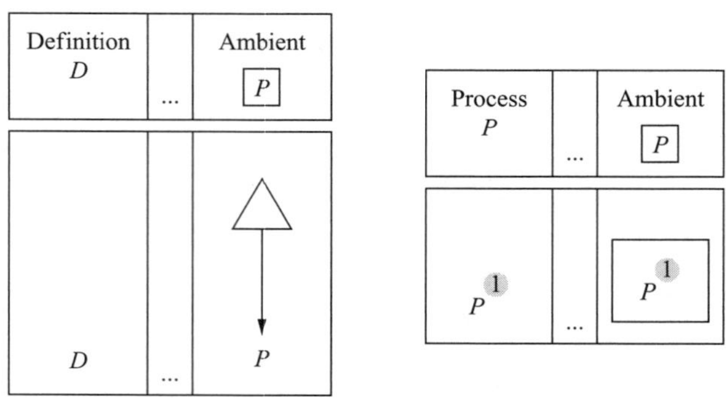

Figure 5.39 Graphical representation for SPiM with compartments. The static representation of an ambient extends the display of static environments E in Figure 5.36, while the dynamic representation extends the display of dynamic processes P in Figure 5.37

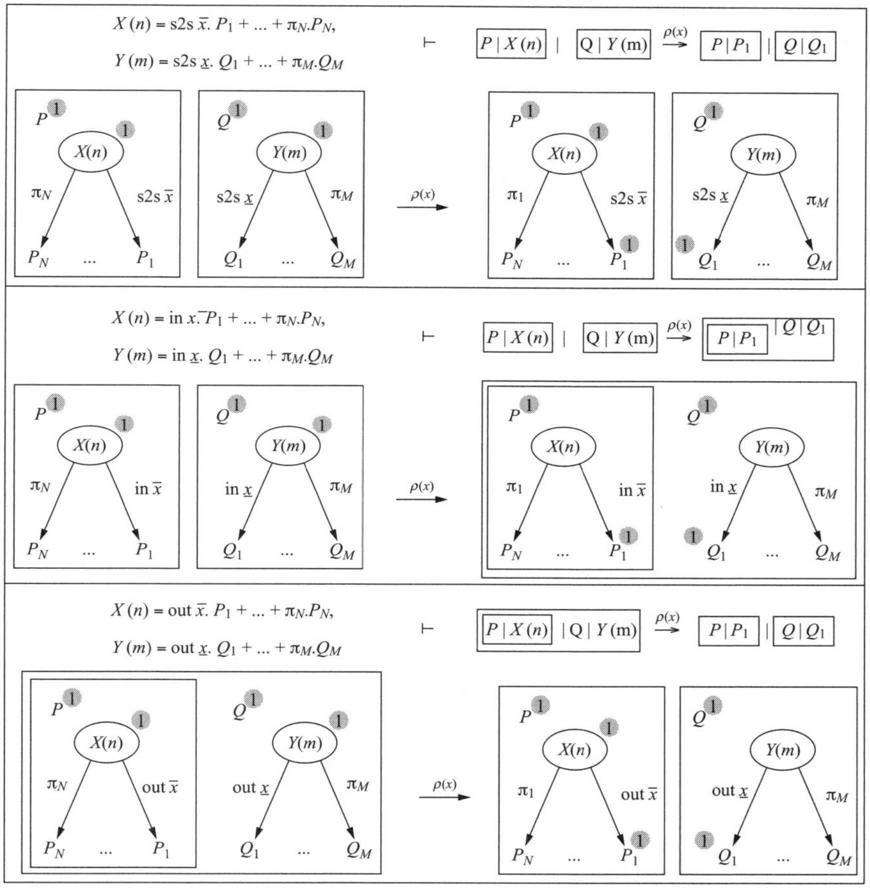

Figure 5.40 Example graphical reactions in SPiM with compartments, where the graphical representation at the bottom is equivalent to the textual representation at the top. The examples illustrate the reaction rules (5.29), (5.30), and (5.31) of Definition 5.8, respectively

Acknowledgments

Thanks to Hillel Kugler for discussions and helpful comments on *C. elegans* modeling and to Rosie Bloxsom for contributing a stochastic VPC model based on [37]. Thanks also to Yaki Setty, Tim Elliott, Alistair Bailey, and Sandra Phillips for additional feedback.

References

[1] R. Blossey, L. Cardelli, and A. Phillips. A compositional approach to the stochastic dynamics of gene networks. *Transactions in Computational Systems Biology*, 3939:99–122, January 2006.

[2] R. Blossey, L. Cardelli, and A. Phillips. Compositionality, stochasticity and cooperativity in dynamic models of gene regulation. *HFSP Journal*, 2(1):17–28, February 2008.

[3] J. Borgstroem, A. Gordon, and A. Phillips. A chart semantics for the pi-calculus. *Electronic Notes in Theoretical Computer Science*, 194(2):3–29, January 2008.

[4] L. Cardelli. Brane calculi. In V. Danos and V. Schächter, editors, *CMSB'04*, pages 257–278, Berlin: Springer, 2004.

[5] L. Cardelli and A. D. Gordon. Mobile ambients. *Theoretical Computer Science*, 240(1):177–213, 2000. An extended abstract appeared in *FoSSaCS '98*: 140–155.

[6] L. Cardelli and G. Zavattaro. On the computational power of biochemistry. In K. Horimoto, G. Reqnesburger, M. Rosenkranze, and H. Yoshida, editors, *Algebraic Biology*, volume 5147 of *Lecture Notes in Computer Science*, pages 65–80. Berlin: Springer, July 2008.

[7] N. Chabrier-Rivier, F. Fages, and S. Soliman. The biochemical abstract machine BIOCHAM. In V. Danos and V. Schächter, editors, *Proceedings of CMSB*, volume 3082 of *Lecture Notes in Computer Science*, pages 172–191. Berlin: Springer, 2004.

[8] F. Ciocchetta and J. Hillston. Bio-pepa: An extension of the process algebra pepa for biochemical networks. *Electronic Notes in Theoretical Computer Science*, 194(3):103–117, 2008.

[9] P. W. Sternberg, C. A. Giurumescu, and A. R. Asthagiri. Intercellular coupling amplifies fate segregation during caenorhabditis elegans vulval development. *Proceedings of the National Academy of Sciences*, 103(5): 1331–1336, January 2006.

[10] V. Danos and C. Laneve. Formal molecular biology. *Theoretical Computer Science*, 325(1):69–110, 2004.

[11] M. B. Elowitz and S. Leibler. A synthetic oscillatory network of transcriptional regulators. *Nature*, 403(6767):335–338, January 2000.

[12] J. Fisher, N. Piterman, A. Hajnal, and T. A. Henzinger. Predictive modeling of signaling crosstalk during *C. elegans* vulval development. *PLoS Computational Biology*, 3(5):0862–0873, May 2007.

[13] D. T. Gillespie. Exact stochastic simulation of coupled chemical reactions. *Journal of Physics and Chemistry*, 81(25):2340–2361, 1977.

[14] C. C. Guet, M. B. Elowitz, W. Wang, and S. Leibler. Combinatorial synthesis of genetic networks. *Science*, 296:1466–1470, May 2002.

[15] M. Hucka et al. The systems biology markup language (SBML): A medium for representation and exchange of biochemical network models. *Bioinformatics*, 19(4):524–531, 2003.

[16] C. Kuttler, C. Lhoussaine, and J. Niehren. A stochastic pi calculus for concurrent objects. In H. Anai, K. Horimoto, and T. Kutsia, editors, *Algebraic Biology*, volume 4545 of *Lecture Notes in Computer Science*, pages 232–246. Berlin: Springer, August 2007.

[17] P. Lecca and C. Priami. Cell cycle control in eukaryotes: A BioSPI model. In *BioConcur'03*. ENTCS, 2003.

[18] R. Milner. *Communicating and Mobile Systems: The π-Calculus*. Cambridge: Cambridge University Press, 1999.

[19] R. Milner, J. Parrow, and D. Walker. A calculus of mobile processes, part I/II. *Journal of Information and Computation*, 100:1–77, September 1992.

[20] K. M. Murphy, P. Travers, and M. Walport. *Janeway's Immunobiology*, 7th ed. London: Garland Science, 2008.

[21] R. Marelly, L. Appleby, J. Fisher, A. Pnueli, D. Harel, M. J. Stern, E. Jane, A. Hubbard, N. Kam, and H. Kugler. A scenario-based approach to modeling development: A prototype model of *C. elegans* vulval fate specification. *Developmental Biology*, 323(1):1–5, November 2008.

[22] D. Noble. *The Music of Life: Biology Beyond the Genome*. Oxford: Oxford University Press, 2006.

[23] M. Pedersen and G. Plotkin. A language for biochemical systems. In M. Heiner and A. M. Uhrmacher, editors, *Proceedings of CMSB*, *Lecture Notes in Computer Science*. Berlin: Springer, 2008.

[24] A. Phillips. An abstract machine for the stochastic bioambient calculus. *Electronic Notes in Theoretical Computer Science*, 227:143–159, January 2009.

[25] A. Phillips and L. Cardelli. Efficient, correct simulation of biological processes in the stochastic pi-calculus. In M. Calder and S. Gilmore, editors, *Computational Methods in Systems Biology*, volume 4695 of *Lecture Notes in Computer Science*, pages 184–199. Berlin: Springer, September 2007.

[26] A. Phillips, L. Cardelli, and G. Castagna. A graphical representation for biological processes in the stochastic pi-calculus. *Transactions in Computational Systems Biology*, 4230:123–152, November 2006.

[27] C. Priami and P. Quaglia. Beta binders for biological interactions. In V. Danos and V. Schächter, editors, *Proceedings of CMSB*, volume 3082 of *Lecture Notes in Bioinformatics*, pages 20–33. Berlin: Springer, 2005.

[28] C. Priami, A. Regev, E. Shapiro, and W. Silverman. Application of a stochastic name-passing calculus to representation and simulation of molecular processes. *Information Processing Letters*, 80:25–31, 2001.

[29] C. Priami. Stochastic π-calculus. *Computer Journal*, 38(6):578–589, 1995.

[30] A. Regev, W. Silverman, and E. Shapiro. Representation and simulation of biochemical processes using the pi-calculus process algebra. *Pacific Symposium on Biocomputing*, 6:459–470, 2001.

[31] A. Regev, E. M. Panina, W. Silverman, L. Cardelli, and E. Y. Shapiro. Bioambients: An abstraction for biological compartments. *Theoretical Computer Science*, 325(1):141–167, 2004.

[32] A. Regev and E. Shapiro. The π-calculus as an abstraction for biomolecular systems. *Modelling in Molecular Biology*, pages 219–266, 2004.

[33] D. Sangiorgi and D. Walker. *The π-Calculus: A Theory of Mobile Processes.* Cambridge: Cambridge University Press, 2001.

[34] L. M. Sompayrac. *How the Immune System Works*, 3rd ed. Malden, MA: Wiley-Blackwell, 2008.

[35] P. W. Sternberg. Vulva development. WormBook, ed. June 2005. The C. elegans Research Community, Wormbook, doi/10.1895/Wormbook.161 http://www.wormbook.org

[36] P. W. Sternberg and H. R. Horvitz. Pattern formation during vulval development in *C. elegans*. *Cell*, 44:761–772, March 1986.

[37] P. W. Sternberg and H. R. Horvitz. The combined action of two intercellular signaling pathways specifies three cell fates during vulval induction in *C. elegans*. *Cell*, 58:679–693, August 1989.

[38] J. W. Yewdell, E. Reits, and J. Neefjes. Making sense of mass destruction: Quantitating MHC class I antigen presentation. *Nature Reviews Immunology*, 3:952–961, 2003.

Chapter

6

Software Tools for Biological Pathway Modeling

Mary F. McGuire and M. Sriram Iyengar

School of Health Information Sciences, University of Texas Health Science Center at Houston, Houston, TX 77030

Abstract Modeling biological pathways requires data sources plus theoretical and applied methods for data display, model building, and checking. This chapter reviews a selection of pathway software tools suitable for translational biomedical research.

6.1 Introduction

Systems biology research has recently spawned an ever-increasing number of methodologies and software tools, dealing with the entire spectrum of investigation, from molecular activity to clinical analysis. "Systems Biology" was introduced in 2005 as a Medical Subject Heading Term (MeSH) for articles indexed for the U.S. National Library of Medicine's MEDLINE/PubMed. It is under the category of "Computational Biology." "Systems Biology" (MeSH) is defined as the comprehensive, methodical analysis of complex biological systems by monitoring responses to perturbations of biological processes and the large-scale, computerized collection and analysis of the data that are used to develop and test models of biological systems. As of 2008, PubMed listed more than 1898 systems biology articles, of which one-third dealt with methods or statistics and numerical data. Google Scholar displayed 393,000 references for systems biology, with approximately one-third relating to methods or statistics and numerical data. In contrast, as of 2008 the IEEE Xplore portal retrieved only 532 systems biology papers, while the ACM portal retrieved 311.

A number of survey papers have been published, listing hundreds of references for systems biology tools. These papers provide overviews, but rarely provide an operational framework of the way in which the tools may be utilized, their underlying theories, and their strengths in particular applications.

Trying to find specific tools for specific projects is difficult, due to the ever-changing scope of systems biology and the lack of cross-disciplinary indexes. For example, the clinical research community looks to PubMed, and computer science to ACM and IEEE (which are rarely indexed in PubMed). Often information about the latest tools and methods can be found only with a well-formed Google search.

A key area of systems biology research is the modeling of biological and biochemical processes as evidenced in metabolic and signaling pathways. When integrated with data from clinical studies, pathway modeling may lead to drug discovery as well as alternative treatments that result in improved patient outcomes.

This chapter highlights some software tools that may provide an operational framework for the study of biological pathway modeling in translational medical research. Prospective users include biomedical mathematicians, systems biologists, and medical researchers.

Discussion of the software tools covers the following Primary Function areas:

1. Data source. Data for pathway nodes (species, such as kinases) and edges (processes, such as protein–protein interactions). Comprehensive data sources for temporal (kinetic and cell cycle phase) and spatial (location within the cell) data were also sought, but not found.
2. Data display. Graphical display (2D and 3D) to visualize the pathway data.
3. Model building. Tools to build pathway models, using techniques both numerical (ordinary differential equations, statistics) and logical (pi-calculus, process algebra). Most models are easily modified for data exploration.
4. Model checking. Tools to verify that the pathway model created was correct.

Note that tools for model building and checking should be based on a solid theoretical foundation.

6.1.1 Criteria

The tools listed here have been in use for some time to model biological pathways, and have been cited for research in signaling and metabolic pathways. Biological pathways may look like computer networks, but their behavior is very different. Although computer network tools may be modified for this use, we chose to deal with only those tools that had already demonstrated some success in this area. Searches were limited to tools that had published pathway results.

Pathway data are in constant flux, as new discoveries flood the journals and online databases daily. Recognizing this fact, we chose to look for a minimum number of reliable data sources to feed our models.

Data display needs to facilitate queries and permit annotations for biological systems, medical, and drug research, including links to internal databases. Visualization should be as intuitive as possible, and handle 2D and 3D displays.

Tools for modeling and modeling checking were preferably based on a solid theoretical foundation, and were proven useful for a variety of biological systems.

All tools were to be currently maintained, and preferably available on a cross-platform basis. User-friendly graphical user interfaces (GUIs) and simple language constructs were preferred.

Tool cost was a consideration. Nevertheless, we prefer to use our resources for pathway research rather than for internal software development and support.

Underlying theoretical methods for pathway modeling tools include Bayesian network modeling, interval probability, semantic web, RDF, Gillespie's stochastic simulation algorithm (SSA), stochastic Petri networks, and probabilistic Boolean networks.

6.1.2 Methods to Find Tools

New methods and tools for systems biology are coming out daily, so the list of tools given here merely provides a sense of where the field is developing. It's recommended that the method of finding these tools be used regularly to locate changes, updates, and new directions.

Systematic searches were used on PubMed, IEEE, ACM, and Google Scholar for publications through December 31, 2008.

The following search strategies were used:

PubMed http://www.ncbi.nlm.nih.gov "Systems Biology"[MeSH]
("Systems Biology/methods"[MeSH] OR "Systems Biology/statistics and numerical data"[MeSH])

IEEE http://ieeexplore.ieee.org (('systems biology')<in>metadata
((('systems biology')<in>metadata) <and> ((method)<in>metadata))
((('systems biology')<in>metadata) <and> ((statistics)<in>metadata)) <and> (('numerical data')<in>metadata) (note: no results)

((('systems biology')<in>metadata) <and> ((statistics)<in>metadata))
((('systems biology')<in>metadata) <and> ((data)<in>metadata))

ACM http://portal.acm.org/portal.cfm ("Systems Biology")
("Systems Biology") and (method)
("Systems Biology") and (statistics)
("Systems Biology") and (data)

Google Scholar http://scholar.google.com "Systems Biology"
"Systems Biology" method
"Systems Biology" statistics
"Systems Biology" data
"Systems Biology" "numerical data"
"Systems Biology" statistics "numerical data"

From these systemic searches for primary sources, additional information was gleaned from references in these articles and correspondence.

6.2 Tools

6.2.1 BIOCHAM

Description: BIOCHAM (Biochemical Abstract Machine) is a programming environment for modeling biochemical systems, running simulations, and querying the model in temporal logic. BIOCHAM manipulates formal objects that represent chemical or biochemical compounds, ranging from small molecules to macromolecules and genes. BIOCHAM objects can also be used to represent control variables and abstract processes. This software can interface to the NuSMV model checker (discussed later in this chapter) for evaluating CTL queries.

Primary Function: Model building

Underlying Method: Temporal logic: CTL or LTL

Language/Interface: Java 1.5

Platforms: Windows, Linux, Mac OS X (10.4 or greater)

Website: http://contraintes.inria.fr/BIOCHAM

Reference: L. Calzone, F. Fages, and S. Soliman BIOCHAM: An environment for modeling biological systems and formalizing experimental knowledge. *Bioinformatics,* 22(14):1805–1807, July 15, 2006.

6.2.2 BioCyc

Description: BioCyc is a collection of more than 400 pathway/genome databases (PDGBs). Each PDGB in the BioCyc collection describes the genome and metabolic pathways of a single organism, with the exception of the MetaCyc database, which is a reference source on metabolic pathways from many organisms. BioCyc includes the databases EcoCyc (*Escherichia coli* K-12 MG1655 Model-Organism Database), and MetaCyc (Multiorganism Metabolic Pathway and Enzyme Database).

Primary Function: Data source

Underlying Method: Database

Language/Interface: BioPAX, Pathway Tools, SBML, Gene Ontology annotations (EcoCyc only)

Platforms: Web: Internet Explorer version 6, Firefox version 2, Safari version 3 or better

Website: http://www.biocyc.org

Reference: R. Caspi, H. Foerster, C. A. Fulcher, P. Kaipa, M. Krummenacker, M. Latendresse, S. Paley, S. Y. Rhee, A. G. Shearer, C. Tissier, T. C. Walk, P. Zhang, and P. D. Karp. The MetaCyc database of metabolic pathways and enzymes and the BioCyc collection of pathway/genome databases. *Nucleic Acids Research,* 36(Database issue):D623–D631, January 2008.

6.2.3 BioLayout Express^{3D}

Description: BioLayout Express^{3D} is a system for layout, visualization, and clustering of large network graphs in 3D and 2D, such as graphs generated from large-scale gene expression studies with Affymetrix chip experiments. In principle, this functionality can be used to generate graphs of other data types whose input consists of unique nodes accompanied by a vector of measurements for that node across different conditions. In the case of Affymetrix data, nodes represent probesets and the vector consists of normalized expression measurements for each probeset across the conditions tested.

Primary Function: Model building

Underlying Methods: Fruchterman–Rheingold layout algorithm for 2D/3D graph positioning and display; Markov Clustering (MCL) algorithm for graph clustering

Language/Interface: Java 1.6

Platforms: Mac OS X, Windows, Linux

Website: http://oriol.gti.ed.ac.uk/BioLayout/index.shtml

Reference: T. C. Freeman, L. Goldovsky, M. Brosch, S. van Dongen, P. Maziere, R. J. Grocock, S. Freilich, J. Thornton, and A. J. Enright. Construction, visualisation, and clustering of transcription networks from microarray expression data. *PLoS Computational Biology,* 3(10):2032–2042, October 2007.

6.2.4 BiologicalNetworks

Description: BiologicalNetworks is an integrated research environment (IRE) for biological pathways analysis, querying, and visualization of gene regulation and protein interaction networks, metabolic pathways, and signaling pathways. The database currently integrates more than 100 curated and publicly contributed data sources for thousands of eukaryotic, prokaryotic, and viral genomes. It is equipped with filtering and visualization tools to provide high-quality, easily understood scientific presentation of pathway analysis results.

Primary Function: Data source, data display, model building

Underlying Method: Database

Language/Interface: Java 1.5

Platforms: Windows 2000 and XP, Linux, and Macintosh

Website: http://biologicalnetworks.net

Reference: M. Baitaluk, M. Sedova, A. Ray, and A. Gupta. BiologicalNetworks: Visualization and analysis tool for systems biology. *Nucleic Acids Research,* 34(Web Server issue):W466–W471, 2006.

6.2.5 BioSPI

Description: BioSPI simulates the behavior of biochemical systems using pi-calculus and ambient calculus. It is based on the Logix system, which implements Flat Concurrent Prolog (FCP). The use of FCP allows both mobility and synchronized communication, two of the major features of pi-calculus. This is believed to be the only implementation of the full synchronous calculus, including mixed choice.

Primary Function: Model building

Underlying Method: Stochastic pi-calculus

Language/Interface: Not given

Platforms: Sun Solaris, SGI, Linux

Website: http://www.wisdom.weizmann.ac.il/~biospi/index_main.html

Reference: A. Regev, W. Silverman, and E. Shapiro. Representation and simulation of biochemical processes using the pi-calculus process algebra. *Proceedings of the Pacific Symposium of Biocomputing 2001,* 6:459–470, 2001.

6.2.6 Bio-SPICE (Biological Simulation Program for Intra- and Inter-cellular Evaluation)

Description: Bio-SPICE is an open-source framework and software toolset for systems biology that assists biological researchers in the modeling and simulation of spatio–temporal processes in living cells. There are more than

80 Bio-SPICE core tools that can be added as plugins to the Bio-SPICE Dashboard.

Primary Function: Model building

Underlying Method: Depends on tool

Language/Interface: Java

Platforms: Windows, Linux, Mac OS X (PowerPC)

Website: http://biospice.sourceforge.net

Reference: T. D. Garvey, P. Lincoln, C. J. Pedersen, D. Martin, and M. Johnson. BioSPICE: Access to the most current computational tools for biologists. *OMICS*, 7(4):411–420, Winter 2003.

6.2.7 BioWarehouse

Description: BioWarehouse is an open-source software environment for integrating a set of biological databases into a single physical database management system for data management, mining, and exploration. It is populated using loader programs that translate the flat file representation of a source database into the warehouse schema. A loader is provided for each source database supported by BioWarehouse. Once loaded within a BioWarehouse instance running on a database such as MySQL, a set of source DBs can be queried together. BioWarehouse is a component of the Bio-SPICE project.

Primary Function: Data source

Underlying Method: Database

Language/Interface: C, Java, Oracle, MySQL, Perl

Platform: Linux

Website: http://biowarehouse.ai.sri.com

Reference: T. J. Lee, Y. Pouliot, V. Wagner, P. Gupta, D. W. Stringer-Calvert, J. D. Tenenbaum, and P. D. Karp. BioWarehouse: A bioinformatics database warehouse toolkit. *BMC Bioinformatics,* 7:170, 2006.

6.2.8 Cellerator

Description: Cellerator is a Mathematica package designed to facilitate biological modeling via automated equation generation. It was designed with the intent of simulating at least the following essential biological processes: (1) signal transduction networks (STNs), (2) cells that are represented by interacting signal transduction networks, and (3) multicellular tissues that are represented by interacting networks of cells that may themselves contain internal STNs. Note that xCellerator (discussed in Section 6.2.38) is the second generation of this software.

Primary Function: Model building

Underlying Method: Algebra, Gillespie's stochastic simulation algorithm

Language/Interface: Mathematica, Sigmoid (GUI)
Platforms: Windows, Mac OS X, Linux
Website: http://www.cellerator.org/
References: B. E. Shapiro, A. Levchenko, E. M. Meyerowitz, B. J. Wold, and E. D. Mjolsness. Cellerator: Extending a computer algebra system to include biochemical arrows for signal transduction simulations. *Bioinformatics,* 19(5):677–678, March 22, 2003.
B. E. Shapiro, A. Voroboyov, J. Murakami, and E. D. Mjolsness. New developments in Cellerator. *Poster presented at the International Conference on Systems Biology,* Yokohama, Japan, October 2006.

6.2.9 COPASI

Description: COPASI is a Complex Pathway Simulator for biochemical reaction networks. It models networks with arbitrary kinetic functions and compartments, species, and global quantities. It automatically converts reaction equations to the appropriate mathematical formalism (ordinary differential equations or reaction propensities).
Primary Function: Model building
Underlying Method: LSODA integrator for deterministic solutions; Gibson-Bruck version of Gillespie's stochastic simulation algorithm; optimization algorithms
Language/Interface: C, C++, SBML
Platforms: Windows, Linux, Mac OS X, Sun Solaris
Website: http://www.copasi.org/tiki-index.php
Reference: S. Hoops, S. Sahle, R. Gauges, C. Lee, J. Pahle, N. Simus, M. Singhal, L. Xu, P. Mendes, and U. Kummer. COPASI: A complex pathway simulator. *Bioinformatics,* 22:3067–3074, 2006.

6.2.10 DEGAS Choreographer

Description: Choreographer is a unified tool for the DEGAS (Design Environments for Global ApplicationS) Project. The main purpose of Choreographer is to provide a bridge between the Unified Modeling Language (UML) and process algebras via a set of extractors and reflectors. Currently three process algebras have been incorporated: PEPA for performance models, LySa for security protocols, and PEPA Nets for mobility.
Primary Function: Model building
Underlying Method: Process algebras, UML
Language/Interface: Java (netBeans IDE)
Platforms: Windows, Linux, Sun Solaris, Mac OS X
Website: http://groups.inf.ed.ac.uk/choreographer/

Reference: M. Buchholtz, S. Gilmore, V. Haenel, and C. Montangero, editors. End-to-end integrated security and performance analysis on the DEGAS Choreographer platform. *Proceedings, Formal Methods 2005*, University of Newcastle upon Tyne, UK, 2005.

6.2.11 DMC

Description: DMC (Deductive Model Checking) is a prototype constraint-based symbolic model checker for infinite-state systems with integer and real data variables.

Primary Function: Model checking

Underlying Method: CTL (Computation Tree Logic)

Language/Interface: SICStus Prolog (SICStus Prolog provides constraint programming with several constraint solvers. See http://www.sics.se/isl/sicstuswww/site/index.html.)

Platforms: Windows 2000/XP/Vista/7, Linux, Solaris, Mac OS X (Intel and PowerPC) with SICStus Prolog

Website: http://www.disi.unige.it/person/DelzannoG/DMC/dmc.html

Reference: G. Delzanno and A. Podelski. Model checking in CLP. *Proceedings of the 6th International Conference on Tools and Algorithms for the Construction and Analysis of Systems* (TACAS '99), volume 1579 of *Lecture Notes on Computer Science*, pages 223–239. Berlin: Springer, 1999.

6.2.12 Dynetica

Description: Dynetica is a user-friendly simulator of dynamic networks for constructing, visualizing, and analyzing kinetic models of biological systems. In addition to generic reaction networks, Dynetica facilitates construction of models of genetic networks, where many reactions consist of gene expression and interactions among gene products. Further, it integrates the capability of conducting both deterministic and stochastic simulations. The user needs to simply specify the components in the system and their interactions; the program automatically generates the ordinary differential equations.

Primary Function: Model building

Underlying Method: Deterministic algorithm, Gillespie's stochastic simulation algorithm

Language/Interface: Java 1.5

Platforms: Windows, UNIX, Mac OS X

Website: http://www.duke.edu/~you/Dynetica_page.htm

Reference: L. You, A. Hoonlor, and J. Yin. Modeling biological systems using Dynetica: A simulator of dynamic networks. *Bioinformatics*, 19(3):435–436, February 12, 2003.

6.2.13 Ingenuity Pathways Analysis

Description: Ingenuity Pathways Analysis (IPA) is an all-in-one web-based application that analyzes and integrates data derived from gene expression, microRNA, and SNP microarrays; metabolomics and proteomics experiments; and small-scale experiments that generate gene lists.

Primary Function: Data source, data display, model building

Underlying Method: Database, IPA network generation algorithm

Language/Interface: Web: Java 1.5 or higher

Platforms: Windows Vista, XP SP2 (Internet Explorer 6.0 or higher; Firefox 1.5 or higher); Macintosh 10.4.11, 10.5.2. (Safari 3.1)

Website: http://www.ingenuity.com/products/pathways_analysis.html

Reference: L. Rybaczyk, A. Rozmiarek, K. Circle, I. Grants, B. Needleman, J. E. Wunderlich, K. Huang, and F. L. Christofi. New bioinformatics approach to analyze gene expressions and signaling pathways reveals unique purine gene dysregulation profiles that distinguish between CD and UC. *Inflammatory Bowel Disease,* 15(7):971–984, February 27, 2009.

6.2.14 Mathematica

Description: Mathematica integrates a numeric and symbolic computational engine, graphics system, programming language, documentation system, and advanced connectivity to other applications. It includes thousands of built-in functions for computation, modeling, visualization, development, and deployment. It also includes genome, protein, and chemical data.

Primary Function: Data source, data display, model building

Underlying Method: Mathematical algorithms, programming, and interfaces

Language/Interface: Java, C, .NET, SQL

Platforms: Windows, Mac OS X, Linux, Sun Solaris

Website: http://www.wolfram.com/products/mathematica/index.html

Reference: C. R. Yang, B. E. Shapiro, E. D. Mjolsness, and G. W. Hatfield. An enzyme mechanism language for the mathematical modeling of metabolic pathways. *Bioinformatics,* 21(6):774–780, March 2005.

6.2.15 MATLAB

Description: MATLAB® is a high-level technical computing language and interactive environment for algorithm development, data visualization, data analysis, and numeric computation.

Primary Function: Model building

Underlying Method: Mathematical functions for linear algebra, statistics, Fourier analysis, filtering, optimization, and numerical integration; 2D and 3D graphics functions for visualizing data

Language/Interface: C and FORTRAN APIs, Windows DLLs, Java, Microsoft Component Object Model (COM) and Microsoft ActiveX technologies, web services, and serial port devices

Platforms: Linux, Mac OS X (Intel), Sun Solaris, Windows

Website: http://www.mathworks.com/products/matlab/

Reference: M. Ullah, H. Schmidt, K. H. Cho, and O. Wolkenhauer. Deterministic modelling and stochastic simulation of biochemical pathways using MATLAB. *Systems Biology (Stevenage)*, 153(2):53–60, March 2006.

6.2.16 Maude

Description: Maude is a high-performance reflective language and system supporting both equational and rewriting logic specification and programming.

Primary Function: Model building

Underlying Method: Rewriting logic (pathway logic)

Language/Interface: C, C++

Platforms: Linux, Mac OS X (Intel and PowerPC), Windows (Cygwin–untested)

Website: http://maude.cs.uiuc.edu

References: M. Clavel, F. Durn, S. Eker, P. Lincoln, N. Mart-Oliet, J. Meseguer, and C. Talcott. All about Maude: A high-performance logical framework: How to specify, program, and verify systems in rewriting logic. Volume 4350 of *Lecture Notes in Computer Science: Programming and Software Engineering.* Berlin: Springer, 2007.

See Chapter 1, Symbolic Systems Biology and Pathway Logic

6.2.17 NuSMV

Description: NuSMV is an open architecture for model checking, which can be reliably used for the verification of industrial designs, as a core for custom verification tools, and as a test bed for formal verification techniques; it can also be applied to other research areas. This program can be used with BIOCHAM.

Primary Function: Model checking

Underlying Method: BDD-based (binary decision diagrams) and SAT-based (propositional/Boolean satisfiability) symbolic model checking

Language/Interface: C

Platforms: UNIX, Windows, Solaris

Website: http://nusmv.irst.itc.it/

Reference: A. Cimatti, E. M. Clarke, E. Giunchiglia, F. Giunchiglia, M. Pistore, M. Roveri, R. Sebastiani, and A. Tacchella. NuSMV 2: An open-source tool for symbolic model checking. *Proceedings of the International Conference on Computer-Aided Verification* (CAV 2002). Copenhagen, Denmark, July 27–31, 2002.

6.2.18 OmniGraffle

Description: Diagramming and drawing tool.
Primary Function: Model building
Underlying Method: Graphics
Language/Interface: Imports and exports Microsoft Visio files (Windows)
Platform: Mac OS X
Website: http://www.omnigroup.com/applications/omnigraffle

6.2.19 PathCase: Pathways Database System (Web)

Description: PathCase is an integrated software system for storing, managing, analyzing, and querying biological pathways at different levels of genetic, molecular, biochemical, and organismal detail. PathCase currently runs on three different data sets: (1) KEGG pathways, (2) BioCyc Human pathways (alpha version), and (3) sample pathways from literature.
Primary Function: Data source, data display
Underlying Method: Database
Language/Interface: Java 1.6
Platforms: Web: Firefox, Internet Explorer, Mozilla
Website: http://nashua.case.edu/pathwaysweb
Reference: B. Elliott, M. Kirac, A. Cakmak, G. Yavas, S. Mayes, E. Cheng, Y. Wang, C. Gupta, G. Ozsoyoglu, and Z. Meral Ozsoyoglu. PathCase: Pathways database system. *Bioinformatics,* 24(21):2526–2533, November 1, 2008.

6.2.20 PathCase: Pathways Database System 2.0 (Desktop)

Description: The PathCase Desktop Client is a desktop application for browsing and interacting with the PathCase database. It includes tree viewers to browse metabolic and signaling pathways at varying levels, as well as edit and query tools.
Primary Function: Data source, data display
Underlying Method: Database
Language/Interface: XML
Platforms: Windows .NET, MDAC; requires Tom Sawyer software.
Website: http://nashua.case.edu/PathwaysWeb/Desktop/

Reference: B. Elliott, M. Kirac, A. Cakmak, G. Yavas, S. Mayes, E. Cheng, Y. Wang, C. Gupta, G. Ozsoyoglu, and Z. Meral Ozsoyoglu. PathCase: Pathways database system. *Bioinformatics,* 24(21):2526–2533, November 1, 2008.

6.2.21 PathwayLab

Description: PathwayLab is a software tool for modeling, analysis, and information management of biochemical pathways.

Primary Function: Model building

Underlying Method: Not given

Language/Interface: Mathematica, MATLAB, SBML, Systems Biology Graphical Notation (SBGN)

Platforms: Windows 2000, XP, or Vista with Microsoft Visio 2003 or later

Website: http://innetics.com/

Reference: Q. Zhang, J. Pi, C. G. Woods, and M. E. Andersen. Phase I to II cross-induction of xenobiotic metabolizing enzymes: A feedforward control mechanism for potential hormetic responses. *Toxicology and Applied Pharmacology,* 237(3):345–356, June 15, 2009.

6.2.22 Pathway Tools

Description: Pathway Tools is a comprehensive symbolic systems biology software system that supports several use cases in bioinformatics and systems biology. It supports creation, editing, querying, visualization, and analysis of pathway/genome databases (PGDBs) and has three components: PathoLogic, for metabolic pathway prediction; Pathway/Genome Navigator; and Editor for PGDBs. It is included with BioCyc.

Primary Function: Model building

Underlying Method: Not given

Language/Interface: BioCyc, Oracle, MySQL

Platforms: Linux, MacOS X 10.4 and 10.5 (Intel), Microsoft Windows NT, 2000, XP, Vista

Website: http://brg.ai.sri.com/ptools

Reference: P. Karp, S. Paley, and P. Romero. The Pathway Tools Software. *Bioinformatics,* 18:S225–S232, 2002.

6.2.23 PEPA

Description: PEPA (Performance Evaluation Process Algebra) is a formal language for modeling distributed systems; it is a stochastic process algebra. PEPA models are constructed through the aggregation of components that perform individual activities or cooperate on shared ones. To each activity is attached an estimate of the rate at which it may be performed. Using such a model, a system designer can determine whether a candidate design meets both the behavioral requirements and the temporal requirements demanded of it. PEPA is supported by several tools such as the PEPA Eclipse plugin, which contains a PEPA editor and performance analyzers. PEPA models can also be processed by PRISM, a probabilistic model checker.

Primary Function: Model building

Underlying Method: Stochastic process algebra

Language/Interface: Java 1.5, Java 1.6

Platforms: Mac OS X, Windows, Linux

Website: http://www.dcs.ed.ac.uk/pepa/tools

References: J. Hillston, editor. Process algebras for quantitative analysis. *Proceedings of the 20th Annual IEEE Symposium on Logic in Computer Science,* 2005.

M. Calder, S. Gilmore, and J. Hillston. Modelling the influence of RKIP on the ERK signalling pathway using the stochastic process algebra PEPA. In C. Priami, A. Ingólfsdóttir, B. Mishra, and H. Riis Nelson, editors. *Transactions on Computer Systems Biology VII,* volume 4230 of *Lecture Notes in Bioinformatics,* pages 1–23, 2006.

6.2.24 PRISM

Description: PRISM is a probabilistic model checker—a tool for formal modeling and analysis of systems that exhibit random or probabilistic behavior. It supports three types of probabilistic models: discrete-time Markov chains (DTMCs), continuous-time Markov chains (CTMCs), and Markov decision processes (MDPs), plus extensions of these models with costs and rewards.

Primary Function: Model building and checking

Underlying Method: Markov models: probabilistic, discrete, and continuous

Language/Interface: Java 1.5, C, C++

Platforms: Linux: 32 bit and 64 bit (x86-64, AMD64, Intel 64, Itanium); Windows: up to and including Vista; Mac OS X: up to and including v10.5 (Leopard); Solaris

Website: http://www.prismmodelchecker.org/

References: M. Kwiatkowska, G. Norman, and D. Parker. PRISM: Probabilistic model checking for performance and reliability analysis. *ACM SIGMETRICS Performance Evaluation Review,* 36(4):40–45, March 2009.

See Chapter 2, Probabilistic Model Checking for Systems Biology

6.2.25 Rhapsody/Play-Engine

These two software packages are used with Harel Statecharts for biological modeling.

Rhapsody Description: IBM Rational Rhapsody is a Statechart-based, model-driven development (MDD) environment for systems and software development of real-time and embedded applications. Code that is generated from the model is just another view of the model, which allows the developer to make changes at the model or source level and have either be updated dynamically. Rhapsody generates complete applications in C/C++ and Ada for 8-, 16-, and 32-bit applications.

Primary Function: Model building and checking

Underlying Method: Harel Statecharts

Language/Interface: C, C++, Ada

Platforms: Windows, Linux

Website: http://modeling.telelogic.com

Reference: D. Harel, Statecharts: A visual formalism for complex systems. *Science of Computer Programming*, 8(3), pp. 231–274, June 1, 1987

Play-Engine Description: This software tool captures reactive behavior in a scenario-based fashion, using the language of Live Sequence Charts (LSCs) and either of two techniques (play-in and play-out). It is used with Rhapsody.

Primary Function: Model building and checking

Underlying Method: Inter-object scenario-based behavior of reactive systems

Language/Interface: Live Sequence Charts (LSCs)

Platform: Windows. Note that this software is available only on CD with the following book: David Harel and Rami Marelly. *Come, Let's Play: Scenario-Based Programming Using LSCs and the Play-Engine,* Berlin: Springer, 2003.

Website: http://www.wisdom.weizmann.ac.il/~playbook

References: A. Sadot, J. Fisher, D. Barak, Y. Admanit, M. J. Stern, E. J. A. Hubbard, and D. Harel. Toward verified biological models. *IEEE/ACM Transactions on Computational Biology and Bioinformatics,* 5(2):223–234, Los Alamitos, CA: IEEE Computer Society Press, 2008.

See Chapter 4, On Statecharts for Biology

6.2.26 Sigmoid

Description: Sigmoid is a database of models of cellular signaling and metabolic pathways. Sigmoid Model Explorer (SME) is a GUI of the Sigmoid Project. This front end provides access to the web services of the pathway databases and Cellerator, the module responsible for simulations.

Primary Function: Data source, data display, model building

Underlying Method: Database

Language/Interface: Java 1.4+, UML, Cellerator

Platforms: Windows, Linux, Mac OS X

Website: http://www.sigmoid.org/

Reference: J. Cheng, L. Scharenbroich, P. Baldi, and E. Mjolsness. Sigmoid: Towards an intelligent, scalable, software infrastructure for pathway bioinformatics and systems biology. *IEEE Intelligent Systems and Their Applications*, 20(3):68–75, May–June 2005.

6.2.27 SimBiology

Description: SimBiology extends MATLAB with tools for modeling, simulating, and analyzing biological systems.

Primary Function: Model building

Underlying Methods: Stochastic and deterministic solvers, ordinary differential equations, optimization

Language/Interface: Microsoft Excel file, text file, MAT-file or the MATLAB workspace, SBML

Platforms: Windows, Linux, Solaris 10, Mac OS X(Intel); requires MATLAB software

Website: http://www.mathworks.com/products/simbiology

Reference: Matthew Schultz. Modeling hypocotyl elongation in Arabidopsis. *Explorations: An Undergraduate Research Journal at UC Davis*, 11:49–66, 2008.

6.2.28 Simpathica

Description: Simpathica is a set of tools to simulate and analyze biological pathways.

Primary Function: Model building

Underlying Method: Semi-algebraic hybrid automata/TCTL (Timed Computation Tree Logic)

Language/Interface: Bio-SPICE, SBML

Platforms: Windows, Linux

Website: http://www.bioinformatics.nyu.edu/Projects/Simpathica

Reference: M. Antoniotti, S. Paxia, N. Ugel, and B. Mishra. Simpathica: A computational system biology tool within the Valis bioinformatics environment. In E. Eiles and A. Kriete, editors. *Computational Systems Biology*, pages 79–102. Burlington: Academic Press, 2006.

6.2.29 Snoopy

Description: Snoopy is a software tool to design and animate hierarchical graphs including Petri nets. It can be used to verify technical systems, especially software-based systems, as well as validate natural systems such as metabolic, signal transduction, and gene regulatory biochemical networks.

Primary Function: Model building and checking

Underlying Method: Petri nets

Language/Interface: SBML, PRISM

Platforms: Windows, Linux, Mac OS X 10.5 (Intel)

Website: http://www-dssz.informatik.tu-cottbus.de/index.html?/software/snoopy.html

References: M. Heiner, R. Richter, C. Rohr, and M. Schwarick. Snoopy: A tool to design and execute graph-based formalisms. [Extended Version]. *Petri Net Newsletter*, 74:8–22, April 2008. This paper is an extended version of *Proceedings of PNTAP 2008* (associated to SIMUTools 2008), Marseille, France, March 2008.

See Chapter 3, Petri Nets for Systems Biology

6.2.30 SPiM

Description: The Stochastic Pi Machine (SPiM) is a simulator for the stochastic pi-calculus that can be used to execute models of biological systems.

Primary Function: Model building

Underlying Method: Stochastic pi-calculus

Language/Interface: Objective Caml

Platforms: Command-line version: Windows, Mac OS X, Linux; graphical interface version: Windows with .NET 2.0

Website: http://research.microsoft.com/~aphillip/spim/

References: A. Phillips and L. Cardelli. Efficient, correct simulation of biological processes in the stochastic pi-calculus. *Proceedings of Computational Methods in Systems Biology (CMSB'07)*, 4695:184–199, September 2007.

See Chapter 5, A Visual Process Calculus for Biology

6.2.31 STODE

Description: STODE automatically parses and performs stochastic simulation of a system of ordinary differential equations.

Primary Function: Model building

Underlying Method: Algebra, Gillespie's stochastic simulation algorithm

Language/Interface: C

Platforms: Windows, Linux

Website: http://projects.eml.org/bcb/software/software/Carel

Reference: C. van Gend and U. Kummer. STODE: Automatic stochastic simulation of systems described by differential equations. *ICSB 2001: 2nd International Conference on Systems Biology*, 2001.

6.2.32 Tom Sawyer

Description: Tom Sawyer software is used for the development of high-performance graph visualization, layout, and analysis systems.

Primary Function: Data display, model building

Underlying Method: Not given

Language/Interface: Microsoft Studio, ActiveX, C++, Java, .NET, Visio, XML

Platform: Windows

Website: http://www.tomsawyer.com/home/index.php

Reference: T. V. Prasad and S. I. Ahson. Visualization of microarray gene expression data. *Bioinformation,* 1(4):141–145, 2006.

6.2.33 TRANSPATH

Description: TRANSPATH provides information on signaling molecules, metabolic enzymes, second messengers, endogenous metabolites, miRNAs, and the reactions they are involved in, resulting in a complex network of interconnected chains of reactions and pathways. TRANSPATH focuses on cross-talk between metabolic and signaling pathways as well as on signaling cascades that change the activities of transcription factors and thus alter the gene expression profile of a given cell. It can be used with Cell Illustrator and CSML to graphically model and simulate cellular processes (see the Nagasaki et al., 2008, reference).

Primary Function: Data source

Underlying Method: Database

Language/Interface: Web

Platforms: Multiplatform

Website: http://www.gene-regulation.com/pub/databases.html

References: M. Krull, S. Pistor, N. Voss, A. Kel, I. Reuter, D. Kronenberg, H. Michael, K. Schwarzer, A. Potapov, C. Choi, O. Kel-Margoulis, and E. Wingender. TRANSPATH: An information resource for storing and visualizing signaling pathways and their pathological aberrations. *Nucleic Acids Research,* 34:D546–D551, 2006.

M. Nagasaki, A. Saito, C. Li, E. Jeong, and S. Miyano. Systematic reconstruction of TRANSPATH data into cell system markup language. *BMC Systems Biology,* 2:53, 2008.

6.2.34 Tulip

Description: The Tulip framework allows the visualization, drawing, and editing of huge 3D graphs having more than 1 million elements. It allows navigation through geometric operations as well as the extraction of subgraphs and the enhancement of the results obtained by filtering.

Primary Function: Data display, model building

Underlying Method: OpenGL rendering

Language/Interface: C, C++

Platforms: Linux, Mac OS X, Windows

Website: http://tulip.labri.fr/news.php

References: D. Archambault, T. Munzer, and D. Auber. Smashing peacocks further: Drawing quasi-trees from bioconnected components. *IEEE Transactions on Visualization and Computer Graphics,* 12(5), September–October 2006.

R. Bourqui, Auber, V. Lacroix, and F. Jourdan. Metabolic network visualization using a constraint planar graph drawing algorithm. *Proceedings of the Information Visualization Conference (IV06),* 2006.

6.2.35 Verus

Description: Verus allows the formal specification and verification of real-time and other time-critical systems. The system being verified is specified in the Verus language and then compiled into a state-transition graph.

Primary Function: Model building and checking

Underlying Methods: Temporal logic including CTL (Computational Tree Logic), RTCTL (Real Time Computational Tree Logic)

Language/Interface: C

Platforms: Linux (primary)

Website: http://www.cs.cmu.edu/~modelcheck/verus.html

Reference: S. Campos, E. Clarke, W. Marrero, and M. Minea. Verus: A tool for quantitative analysis of finite-state real-time systems. *Workshop on Languages, Compilers and Tools for Real-Time Systems*, 1995.

6.2.36 Visio 2007

Description: Visio is a diagramming and drawing tool that uses vector graphics to create diagrams.

Primary Function: Model building

Underlying Method: Vector graphics

Language/Interface: Import/export with Microsoft Office; OmniGraffle (Mac), Tom Sawyer, and others

Platform: Windows
Website: http://office.microsoft.com/en-us/visio/default.aspx

6.2.37 xCellerator

Description: xCellerator assists biological modeling via the automated conversion of chemical reactions into ordinary differential equations for signal transduction modeling. xCellerator is backward compatible with the reaction syntax defined in Cellerator, but uses a completely different algorithm to perform the translation.

Primary Function: Model building

Underlying Methods: Algebra, Gillespie's stochastic simulation algorithm

Language/Interface: Mathematica

Platforms: Windows, Mac OS X, Linux, Sun Solaris

Website: http://www.xcellerator.info

Reference: T. S. Najdi, C. R. Yang, B. E. Shapiro, G. W. Hatfield, and E. D. Mjolsness. Application of a generalized MWC model for the mathematical simulation of metabolic pathways regulated by allosteric enzymes. *Journal of Bioinformatics, Computers, and Biology,* 4(2):335–355, April 2006.

6.2.38 Ymer

Description: Ymer is a statistical tool for verifying probabilistic transient properties of continuous-time Markov chains (CTMCs) and generalized semi-Markov processes (GSMPs). Properties are expressed using the Continuous Stochastic Logic (CSL). Ymer includes the hybrid engine from the PRISM tool for CTMC model checking.

Primary Function: Model checking

Underlying Methods: Statistical techniques, based on discrete event simulation and sequential acceptance sampling

Language/Interface: C, C++

Platforms: Multiplatform

Website: http://www.tempastic.org/ymer/

References: H. L. S. Younes, M. Kwiatkowska, G. Norman, and D. Parker. Numerical vs. statistical probabilistic model checking. *International Journal on Software Tools for Technology Transfer*, 8(3):216–228, 2005.

H. L. S. Younes. Ymer: A statistical model checker. In K. Etessami and S. Rajamani, editors, *Proceedings of the 17th International Conference on Computer Aided Verification,* volume 3576 of *Lecture Notes in Computer Science,* pages 429–433. Berlin: Springer, 2006.

6.3 Additional Resources

With the ever-changing field, the best resources seem to be websites with links.

BioUML: A Java framework for systems biology.
http://www.biouml.org/

Portal site for systems biology
http://systems-biology.org/

Systems Biology Markup Language (SBML)
http://sbml.org/Main_Page

Systems Biology Tools and Resources
http://garnet.arabidopsis.org.uk/systems_biology_tools.htm

Systems Biology Workbench Project
http://www.sys-bio.org/research/sbwintro.htm

The Virtual Library of Biochemistry, Molecular Biology, and Cell Biology
http://www.biochemweb.org/systems.shtml

Index